INSIDE THE
MACHINE

An engineer's tale of the modern
automotive industry

MORE GREAT BOOKS FROM VELOCE:

www.veloce.co.uk

First published in hardback in March 2022; colour paperback edition printed June 2022, this mono edition published July 2023 by Veloce Publishing Limited, Veloce House, Parkway Farm Business Park, Middle Farm Way, Poundbury, Dorchester DT1 3AR, England. Tel +44 (0)1305 260068 / Fax 01305 250479 / e-mail info@veloce.co.uk / web www.veloce.co.uk or www. velocebooks.com.
ISBN: 978-1-787119-39-0; UPC: 6-36847-01939-6.

INSIDE THE MACHINE

An engineer's tale of the modern
automotive industry

David Twohig

VELOCE PUBLISHING
THE PUBLISHER OF FINE AUTOMOTIVE BOOKS

Contents

Boys from Cork don't design cars

I have a theory about child development. Most young kids are, at some stage, given a present of a ball of some sort and a model car – a miniature Dinky or Matchbox-type toy car. The kid that reaches out a chubby paw for the ball will probably grow up to be a sportsman or woman, interested in football, rugby, baseball or whatever. But a significant minority will reach for the toy car, and thus spark a lifetime's interest in things mechanical, be they cars, trains, boats, or planes – possibly all of these things.

My own chubby paw very definitely reached out for the toy car. Being born in 1970, it was probably a 1:43 scale Ford Consul or Zephyr, or some other 1960s hulk of that type. Grainy 8mm home cinema shot by my father shows me in nappies, enthusiastically pushing around a tricycle, or zooming around in a rather natty pedal car, often with a tiny toy car clutched in one hand, just in case.

My hometown is Cork, in the south-west of the Republic of Ireland, a small country with little automotive engineering heritage. However, I had the great fortune to have a father who was a true car enthusiast, a gifted self-taught mechanic and a natural DIY engineer. As soon as I could hold a spanner I was recruited into various mechanical jobs. Dad was always repairing, and later restoring, various cars of very mixed pedigrees. Typical of his wartime generation, he was very averse to spending money, least of all on professional mechanics, most of whom he held in rather low regard. There was no alternative to DIY, and no limit to the complexity of tasks that he would attempt, from full engine rebuilds, to bodywork repairs, to spray painting. Even punctures on relatively modern cars were repaired at home, popping the tyre beads off the wheel rim in a huge wobbly vice; a method that would make modern health and safety inspectors blanch. I had an idyllic childhood, messing around with greasy fingers and learning by watching my father. Later, in my teens, my father took early retirement, and spent his spare time restoring a 1926 Ford Model T, literally, and with none of the usual exaggeration implied by this over-used phrase, from a rusty pile of parts. This labour of love took several years, and my memories of school and early university years are of taking breaks from study to help him resurrect this icon of automotive and industrial engineering to its former glory: sanding rust off 60-year-old chrome-vanadium steel, seeing the famous oval with that baroque Ford Motor Company script surfacing from the grip of the red oxides that had seized it decades before. This taught me how cars – albeit very simple cars – work. I also learned that they were far more than mere machines to transport people from A to B – that they had an industrial and a human history, that they evoked passion and dedication that went far beyond what anyone ever felt, or could feel, for equally marvellous but ultimately soulless machines like televisions or mobile phones. I absorbed car culture, and probably far too much grease and other hydrocarbons for the sake of my health, at my father's elbow in his garage-cum-workshop. The day I turned 17 I was taught to drive – by my father, of course – at the wheel of his Nissan Bluebird. But I was already in love with cars – becoming old enough to drive them was just the consummation of a life-long passion.

Behind this messing about with cars was, however, a more serious influence from both my parents. Both were from humble origins, having worked hard all their lives to provide their five kids with a comfortable middle-class home and a good education. And they put a *lot* of emphasis on this latter. Again, typical of their generation and background, progression in life was simple in their view – to avoid poverty, you had to have a 'good' job. To have a good job, you had to have a good education. A 'good' education meant a degree – something that only the rich could afford when my parents were young. So, ever since I can remember, school – and afterwards, university – was a very serious business indeed. There was never any outward or overt pressure, but one was given to understand from the age of four that academic failure was simply not an option. I attended the local primary school, and was then sent to the secondary school that they considered would offer me the best chance to get into university – Presentation Brothers' College, Cork. There were three religions in this school – rugby, winning entrance scholarships each year to the local university, and Roman Catholicism. In that order. Too small and too uncoordinated to be any use on a rugby field, and at no time hearing any God(s) calling on me to be a priest, fortunately, I found that I was reasonably bright academically and had a facility for memorising things. Hence, I could at least have a decent shot at the second of the three school religions.

There was no question but to go to university, and not too much hesitation in the choice of courses to pursue, either. My local university – University College, Cork or UCC – was only a few kilometres away, and it offered a prestigious course in Electrical and Electronic Engineering. I was lucky enough to win a scholarship to pay the university fees, which took some financial pressure off my parents, and duly rode my trusty bicycle to and from lectures for the next four years, following a classic, rather old-fashioned, very academic, but extremely well-taught course in the basic engineering building blocks.

Come our final year, most of my classmates took jobs either in software, in management consultancy, or in microelectronics – integrated circuit or silicon chip design. I was very much the odd one out – the only one who wanted a job in an old-fashioned grimy industry that made boring industrial stuff like cars. I spent hours in the university career service offices, pestering the staff there for information, addresses or recruitment brochures for anything and everything to do with making cars. We are still pre-internet, remember – at least in Cork we were. My quest seemed fruitless – it was now mid-1992, and the car industry was starting a worldwide slide into losses, shrinking sales volumes and overcapacity. But sitting rather incongruously on one shelf of the little career services library was a slim recruitment brochure that somehow caught my eye. It was a thin booklet advertising jobs in something called 'Nissan Technical Centre Europe' or NTCE[1]. This explained, in ten or so glossy pages, that Nissan, the second largest Japanese automotive company, had recently created (in 1988, only four years before) a new technology centre in Cranfield in the UK, with the specific vocation of designing a new generation of Nissan models to be built in a state-of-the-art factory in Sunderland in the north of England, and in Nissan's second European production plant, in Barcelona, Spain.

The pages of the brochure were full of standard HR guff – which I naively drank in like an alcoholic still bruised after his fall from the wagon – but what really hooked me was the centre-spread photograph. Ten young lads – and they were all lads, no women – were standing and sitting around, in various uncomfortably stilted poses, like the country cousins at a wedding, in front of a full-size layout drawing of a car – a Nissan Bluebird. The thing that struck me was what they were wearing – neither suits and ties, nor jeans and T-shirts. They were all wearing what I would later come to know were called 'Nissan Blues' or less respectfully 'Smurf suits' – a semi-military suit of appallingly bright blue synthetic fabric, consisting of a pair of straight-up-and down trousers with the front seam sewn in to keep it permanently sharp, and a short military blouson jacket, with two front chest pockets, a name-tag sewn over the left breast pocket, and the Nissan crest – the red and blue 'hamburger' – sewn on as a shoulder flash. This was the standard uniform worn by all production-line workers in Nissan, so it was made of an indestructible heavy-gauge nylon cloth to protect them in the harsh factory environment. All closures were plastic or Velcro to avoid scratching the paint when assembling or repairing cars. But the unique part – the part that struck me – was that it was worn by *everybody*, not only the line-workers. According to the recruitment brochure – and later I was to learn that it was almost true in practice, at least in Nissan's European operations – everybody, from

---

[1] *Nissan's European R&D base was originally called Nissan European Technology Centre (NETC) but changed name in 2000 to become Nissan European Technical Centre (NTCE), a more logical name given that the 'global' R&D centre in Japan was always known as Nissan Technical Center (note American English spelling) or NTC. I will use NTCE throughout this book for simplicity.*

the Managing Director right down to the most junior apprentice, wore this hideous bright blue production worker's suit. The purpose was to underline that it was a single-status company, that there were no boundaries between suit-wearing office workers and blue-collar production workers on the line. Everybody was blue collar – quite literally. The ten young men in the photo were last year's graduate intake. I studied the photo for hours, looking at these guys, maybe a year or so older than me, in their faintly ridiculous but strangely appealing factory workers' garb, smiling self-consciously at each other. I envied them, and was determined that I would do all that I could to get into this strange new company with big ambitions and such terrible fashion sense.

But getting a job there seemed impossible to achieve for a lad from Cork – like dreaming of being a fighter pilot or an astronaut. The number of places available was so tiny, and the number of candidates so large, especially from the UK, which had a long tradition of supplying armies of engineers to the car industry, that it seemed impossible. I sensibly set my sights lower and applied for jobs in other companies, having posted off my oh-so-carefully-filled-in application form to Nissan. I was lucky enough to get some interviews, which I duly attended, trying my best to be enthusiastic about them. Then came a letter in the post marked with the blue and red Nissan 'hamburger'. I tore it open eagerly to find out that I had been invited to a first interview.

A few weeks later I met my very first Nissan employees in a hotel in Dublin, and was pleasantly surprised to find that one of them was a fellow countryman called Stephen Kennedy. So at least one other Irishman *had* made it into this company that intrigued me so much. His colleague was the Human Resources Manager Simon Bottomley, who had commissioned the very recruitment brochure that had so caught my eye. I was terribly nervous, and seriously thought that I had blown the interview. But I had succeeded in making the interviewers laugh, and hoped that at least they might remember me from the other pale-faced, serious and stiff-suited young men that I had met in the lobby of the hotel. A few weeks after, another letter, more stress as I fumbled it open. I had been called to a second, and final interview, in NTCE's Cranfield base! The weeks dragged by until the date when I flew to London and took a train the hour's journey north to the nearest big town to Cranfield – Milton Keynes. I hailed the very first taxi of my life to drive me from the Milton Keynes train station to Nissan's site. I was nervous but excited as the taxi wound its way through rolling picturesque green Buckinghamshire farmland, not at all the landscape I expected.

My first view of Nissan's offices was suitably dramatic – I suddenly spotted a pure white block of buildings rising out of the landscape, on the crest of a long, low ridge of green fields and hedgerows, with dairy cows grazing peacefully here and there. As we got closer, I could see that it was a small but ultra-modern industrial park – a low-rise two-storey office block, built of white-coated steel panels with large glass areas, behind it an intriguingly windowless low-slung white block of a building – presumably too secret to allow prying eyes to look in. The buildings were obviously brand new – in fact, they had been completed only the year before. The few acres of landscaped grounds were still raw, with new trees sprouting here and there, the Japanese-style artificial lake still cloudy. I paid off the taxi with a few of my precious pound notes and pushed open the revolving door of the visitor's reception to the building where I would spend a very large part of the next 13 years of my life.

The next couple of days convinced me even more that I simply had to work here. I fell in love with the place and the company almost at first sight. The people I met were friendly, professional, open, and many of them seemed not much older than myself. The interviews, aptitude tests and group exercises that I participated in along with the 20 or so other candidates seemed practical, well-designed and fuss-free. I bombarded the people I met with questions and absorbed information like the proverbial sponge. We were taken on tours of the shiny new buildings in their incongruously pastoral setting. We wandered through the open-plan offices, another visible symbol of the famous single-status – no closed-away offices here, not even for the Managing Director. I saw my first CAD (Computer Aided Design) stations or 'tubes' with operators staring at giant screens, manipulating mysterious multi-colour 3D models of parts of cars. We had a brief tour of the tantalisingly windowless Test Building. This was, and probably still is, the enthusiast's dream garage – spotless epoxy-painted floors, surgically clean, perfectly lit, with neat rows of four-post lift systems, compressed air hoses spiralling from the ceilings, and immaculate giant Snap-on tool cabinets. Impressively expensive-looking and very new test equipment lurked everywhere – shower spray booths, two- and four-wheel rolling roads, electromagnetically screened rooms for testing radio-wave reception and emission behaviour, and, most exotically, a semi-anechoic sound-proofed chamber with a rolling road where you could (literally) hear your own heart beating if you stayed long enough for your ears to adjust to the spooky total silence. Several Nissan Primera and Micra test cars stood around with technicians working on them, and in various parking bays and chambers I saw carefully covered forms of mysterious vehicles that were obviously not Primeras – I tried hard to hide my excitement at getting a glimpse of these 'secret' cars. In fact, all of this was absolutely normal for any test workshop in any car company, but to an over-enthusiastic and very uncynical fresh graduate engineer from Ireland, it was heaven on earth. At lunchtime we strolled around the grounds and spotted small koi carp swimming in the ornamental lake, a gift from one of the supplier companies. I would see some of these very same carp grow into 2ft (60cm) long monsters in the years to come.

I left Cranfield and flew back home to Ireland having fallen hook, line and sinker for what I had seen and heard. I thought that I had done okay in the interviews and assessment centres, despite the fact that all the other candidates had at first seemed to me to be '*Übermensch*' – tall, good-looking and armed with impressive qualifications from prestigious British universities, and lacking my then still-broad Irish accent. But Nissan had told us that they had 200 candidates for every post, and that they would only be recruiting five people this year – so realistically the chances were remote. In the meantime, I had had a job offer from Ford – which pleased my father, a lifetime Blue Oval man, no end. So, when yet another letter arrived with the now-familiar Nissan logo – this time with an offer of a job as a lowest-rung-of-the-ladder Graduate Engineer, I had to choose between the two companies.

As it turned out, the decision was partly made for me. Simon Bottomley from Nissan HR telephoned to follow up the letter. I happened to be out at the time, and he spoke to my father instead, who duly took a message for me. It was my father – the Ford man – who made my mind up for me, saying "Well, it's your choice, but those lads at Nissan *do* seem to want you badly. And that fella Simon seemed a straight talker." As usual, he was right. I called Simon back to tell him that I would be accepting the job. For the rest of that summer, I enjoyed doing nothing much except spending time with my family. My parents were disappointed that I would be emigrating, but for me it was a logical conclusion. I had known since I was a kid that designing cars was what I wanted to do. There was no automotive industry in Ireland, so it was obvious that one day I would have to leave my native country. And leaving Ireland to work in England was a traditional and accepted thing to do, a path taken by millions before me, including two of my older sisters. So, I left Ireland in September 1992, with one ruck-sack full of clothes, including my one suit and one pair of 'work' shoes, never to return except for short visits, and with absolutely no regrets. My dream job awaited me.

CHAPTER 1

Fresh meat

Nissan Europe in 1992 was an exciting place to be for a newly-minted engineer. Lots was happening, and fortunately I did not know that we were heading into a disastrous period of Nissan's history, that would lead us to the brink of bankruptcy. Much more of that later.

In my first weeks, I learned a lot about Nissan, Nissan Europe, and the particular bit of Nissan Europe I now worked for – Nissan Technical Centre Europe or NTCE. In the early 1980s, Nissan's top management in Japan had decided to 'go global' and to manufacture cars locally in its major overseas markets – first in North America, by far the largest market outside Japan, then in Europe. A state-of-the-art greenfield car plant was to be built in Sunderland, in North East England. The factory was opened in September 1986. The choice of site was highly political. Nissan would probably have preferred to invest in mainland or continental Europe, as most of the component suppliers were to be found in Germany and France, and the vast majority of its sales were in left-hand-drive countries. However, the British government of the day – led by the Iron Lady herself, Margaret Thatcher – welcomed this inward investment from a major Japanese industrial giant. The British 'indigenous' car industry, led by Rover and Ford, was in serious difficulty. The northeast of England was a region particularly hard-hit by the relatively recent loss of its traditional industries. The twin cities of Newcastle and Sunderland, sitting on the Tyne and Wear rivers, had become ghost-towns of derelict shipyards and deserted coal-mine pit-heads. Hence a proposal from a cash-rich Japanese carmaker to build a huge new production plant, which would eventually employ thousands of workers, was a godsend to the beleaguered Conservative Government. A prime site was found near the city of Sunderland, ground was broken in 1984 and in short order the factory – called Nissan Motor Manufacturing UK or NMUK – was rolling, turning out lightly-Europeanised copies of the Nissan Bluebird or 'T72' in Nissan's model code system, the very car in which my father taught me to drive.

At around the same time, Nissan invested heavily in revamping another European factory, Nissan Motor Ibérica SA or NMISA, in Barcelona. This factory had a long history, with roots going back to the 1920s as a tractor and truck manufacturing plant. Nissan had acquired it in 1980, and refurbished it extensively with the intention of becoming the company's specialist plant for 4x4 and commercial vehicles[2].

But Nissan's ambitions in the early and mid-eighties did not stop at simply manufacturing. Nissan planned not only to build, but to *design* cars in Europe – the tag-line at the time was 'designed in Europe, by Europeans, for Europeans.' Nissan was – and still is – the only Japanese carmaker to seriously invest in overseas engineering. Certainly, Honda and Toyota have overseas quality, purchasing, styling and even some engineering capability, but their European products are still very much 'designed in Japan.' Nissan had a very much more decentralised, outward-facing attitude. Hence the establishment of NTCE in 1988, to concentrate on engineering those products to be built in Europe, in the Sunderland and Barcelona plants.

NTCE's first project was the successful 4x4 Terrano II (internal code R20), also built for Ford under licence with the model name Maverick. The vehicle was built in the NMISA plant. Terrano II was launched in 1993 and was to enjoy a very long production run, not being phased out until 2006. Thanks to a simple concept, chunky Tonka-toy styling, and no-frills mechanicals, it was a considerable sales success, although the Nissan version sold better than its Ford-badged sister car. The vehicle was also – crucially – profitable. NTCE's first project was a hit.

In my first weeks and months I also started to learn more about how Nissan had established a truly unique management structure in Europe. Nissan had hired several senior executives from Ford to establish its new European operations – notably Ian (later Sir Ian) Gibson and Peter Wickens. They had the golden opportunity of being able to establish a company structure, organisation and even culture from almost a blank sheet of paper – and they seized it with both hands. Gibson and Wickens blended the best practice production processes and structures that they had observed in the Nissan plants in Japan with key ingredients of European people-management. Somehow, they also succeeded in negotiating a unique deal with the Amalgamated Engineering Union. The British car industry had been brought to its knees by industrial unrest and disastrous management-trade unions relations in the 1970s; for Nissan in the 1980s, Gibson and Wickens proposed a new structure intended to narrow the gaps and reduce the inequality between management and workers. The much-vaunted single-status – and hence the adoption of the (in)famous Nissan Blues staff uniform – was a key part of that. From the Chairman, Ian Gibson himself, to the newest raw recruit fitting parts on the production line, the Terms and Conditions of employment were exactly the same. Everybody had the same daily expenses, healthcare, holidays and working hours. Great emphasis was laid on the adoption of Japanese tools like *Kaizen* (continuous

---

[2] *Sadly, NMISA would close its doors in May 2020, a victim of Nissan's latest financial difficulties.*

improvement), JIT (Just In Time) and *Douki Seisan* (simultaneous ordering). This so-called 'Nissan Management Way' was later to be much studied by business management schools, and much copied. It is easy with hindsight to be somewhat cynical. What was presented as 'this is how it's done in Japan' was in fact quite different from the reality in a Japanese car plant, where things were often far more hierarchical and 'top-down' than portrayed in Europe. But it worked – Nissan was very proud to have never lost even a single minute's production time by strike or industrial unrest of any type. It continues to be a model employer and the winner of dozens of independent awards, not only for productivity and quality, but also for training and career development of its staff.

The flagship model in Europe when I joined the company in 1992 was the Primera Mark I (code P10). This had replaced the Bluebird in production in NMUK in 1990, and was Nissan's offer in the then still crucial family sedan/hatchback D-segment. A close copy of the Japanese domestic market model of the same name, it was a sales success and was reasonably profitable.

Launched the very same month as I joined Nissan, in September 1992, the Micra (code K11) was the second mainstay of NMUK production. It would also enjoy a long and successful production run. It competed in the B-segment or so-called 'supermini' market. It was an innovative little thing, with its radical-in-1992 upturned bathtub styling, 16v DOHC engine and love-it-or-hate-it CVT (Continuously Variable Transmission) automatic gearbox. It even had the dubious honour of winning European Car of the Year in 1993. It was, however, marginally profitable – in the early '90s it was already starting to become difficult to make a profit on small cars built in Europe, and the Micra would dip in and out of the red throughout its lifetime.

My early days at Nissan revolved around these three cars – Terrano II, Primera and Micra. Together with the four other 1992 graduate trainees (two of whom are still at Nissan at the time of writing), we had a first week of CAD training and induction lectures, during which we tried to memorise facts, figures and the endless series of cryptic alpha-numeric codes of which the car industry is so fond. I remember one of the many lectures we received with particular clarity – delivered by a tall, gangly and very articulate Irish engineer (I was to find that I was far from the only Irishman in Nissan) called Dave Kelly. Dave was a Senior Engineer from Body Design, and spoke in tongues for most of his one-hour talk to the five earnest young lads sitting in front of him, but one of the questions he asked stuck with me:

"So, what do you lads think we produce here in NTCE, then?" – a seemingly innocent question delivered in his easy-going Midlands brogue.

"Errm, cars?" ventured one of my fellow trainees.

"Nope, they do that up in NMUK, not here. Did ye see a factory or a great big car park full of cars on the way here? Not cars."

Silence.

"Well, designs for cars, maybe?" ventured another lad.

Approving nod from Dave. "Correct. We produce design data for cars, not cars. That means that we really produce just two things. First, drawings – of parts and of cars. Second, structured lists of which parts you put together to make those cars. Stacks of paper and lists of electronic data. We produce *information* – and that's all. And the lads in Sunderland and in Barcelona have nothing else to go on but our information. So we'd bloody well better get it right, don't you think?"

It was only years later that I realised how important this point was. Dave was helping us to intellectually separate the 'design' process and the production process, in a few clear, unpretentious and deceptively simple sentences. It was a lesson I retained. I was to have the privilege of working with Dave throughout my time in Nissan, and he was one of several role models, right from that first day.

Rather like new recruits in the army, we also received our Nissan uniforms in those first days. I was given two sets of those hideous nylon 'Blues,' with my name embroidered on the breast of the jacket. I can honestly say that I wore it with pride for the next 13 years. I still have one hanging in my wardrobe somewhere – apparently, I'm too sentimental to throw it out.

After these induction preliminaries, we were dispatched to our first 'placements' in an actual engineering section. Nissan's graduate training method at the time was based on not assigning graduates to a destination department until one year or so after recruitment. This was to allow a mixture of formal training courses and short training placements in several different departments to see where each person might best fit. My initial assignment – at my own request – was in Chassis Design. My first, unglamorous, task was to design a mounting for an ABS sensor in the drum-brake version of the Nissan Primera rear axle. Until then, ABS was offered only on all-disc brake versions of Primera.

I dived enthusiastically into my first-ever package study, using my brand-new CAD skills, made the tolerance stack-up calculations – by hand – to verify the clearances between moving parts, and triple-checked the cost-saving calculations. In fact, it was a very standard cost reduction exercise, a trivial engineering task. But to me it was a romantic crusade – I was actually going to design something that did not exist, and if I did it well, it *would* exist, would go into production and be sold in hundreds of thousands. I immersed myself in it, checking and re-checking my calculations, making sure that I had taken into account the dimensional tolerances of each and every part in the chain, not forgetting details like thermal expansion of the various metals involved. Finally, I produced a hand-written technical report (computers were still rare – one clunky IBM PC shared between six engineers) which did, in fact, confirm our engineering hunch: the slightly more compact dimensions of the European (Bosch) wheel speed sensor compared to the

Japanese (Hitachi) original *would* – just – allow it to be packaged in a drum brake assembly. The technical report was duly signed off and in fact ABS *was* to be offered on drum-brake Primeras – the very first part I influenced on a production car.

I also produced my first Design Note during this period. The Design Note was – and still is, nowadays in purely electronic form, of course – the formal document authorising a change to a part. My first such change was very modest indeed. It was for the minor facelift version of Primera, and was again an utterly standard, banal modification. During the pre-production trials of this slightly updated version of the car, the assembly engineers in NMUK had found a potential interference between the front ABS sensor cable and its mounting bracket. They had duly raised a 'concern report' requesting a Design change to solve the problem. Normal business – one of hundreds of such minor tuning issues that are found in such production trials. The only item of note was that it was a safety-critical, or so-called 'Vital' part – anything to do with braking or ABS was automatically flagged up as needing reporting and sign-off at General Manager level. I was assigned the task of investigating the issue. It was not much more than a day's work to confirm that in the worst-case there was indeed an interference between the cable and the little L-shaped pressed-steel bracket that held it to the front suspension strut tower. No problem – the bracket was a new part, and the supplier confirmed that they could still modify the stamping tool to trim a little off the edge of it, without affecting its rigidity. The countermeasure was obvious – cut a bit off the bracket. The more experienced engineer supervising me agreed, saying "Okay, Dave, you decided the change, so you should do the Design Note. It's not very interesting, but you've got to start somewhere. I'll show you how."

He duly talked me through Nissan's already rather-archaic engineering release system, how to change the part number, the part drawing index, specify to which vehicle the change should apply, and from what date. This would directly update the massively complex 'Bill of Materials' system that NMUK used to purchase and sequence parts smoothly onto the production line. All this was summarised on a single piece of paper – the 'Design Note' itself. The drawing of the part was also updated to show the modified dimensions. I signed the various documents in the boxes marked 'Designer' for the first time before taking it to the Department Manager for his approval. This latter didn't even look up as he scribbled his initials in the appropriate box, just saying "This one's a Vital Part as you know Dave, so it needs General Manager approval as well. So just pop over to take the Boss through it, will you?"

Gulp. 'The Boss' of Vehicle Engineering at this time was a gentleman called Shingyoji-san, and even a newbie like me knew his daunting reputation for not suffering fools gladly. Clutching my stack of paperwork, and a sample part of the offending bracket, I set off for Shingyoji-san's lair. I walked nervously up to his desk, coughed discreetly and said "Shingyoji-san? Excuse me. Could I please explain a Design Note that needs your signature?"

The Boss looked up – I noticed with relief that there was no flame or smoke coming out of his nostrils or any other orifice as far as I could tell – and motioned me to take a seat. I stammered through my carefully rehearsed explanation, waving the sample part around like a talisman as he casually leafed through the paperwork with a practised air. "Hmmm. So, tool can still be modified in time for the next trial production build?" Yes, I assured him, noting that his eye fell for a moment on the modest tooling modification cost. Without a further word, he scratched a signature in the required 'General Manager' box, thanked me politely and turned back to his own work to indicate that I was dismissed. I stood there almost disappointed ... I had clearly been panicking for nothing. I came back to the Chassis team who were waiting for me with smiles on their faces – it had been another test. But I did not mind their laughter at my expense – I had actually modified a real part on a real car, to be sold to real customers. I was at last, a car design engineer.

Another crucial part of the newcomer's induction was the week of 'line training' that all new company employees underwent, working on the production line at NMUK. This was a truly humbling experience, and anyone who has actually worked in an efficient car plant cannot come away without immense respect for those who earn a living in this genuinely hard environment. We did a different job every day – the simplest and most repetitive jobs, partly because the lads on the line understandably enjoyed giving shiny new graduates the most boring tasks available, and partly because we were physically incapable of doing the more complicated jobs. I spent one day fitting exhausts, another getting under the feet of one of the guys operating one of the giant 5000 tonne presses in the cavernous Press Shop that stamped out the in-house sheet metal parts like roofs, doors and body sides. I fitted core blanking plugs into the sides of engine blocks, and ripped my thumbs to shreds clipping in electrical harnesses. It was a relief to go back to our cushy 'office' jobs, a lot humbler and hopefully a little wiser.

For the final period of my training, I was assigned to Trial Production. This was the department that built prototype vehicles, and they needed a junior engineer to look after quality assurance of the incoming prototype parts. An interesting enough job, but not for me – I wanted to design things, and quickly grew frustrated. Fortunately, I soon heard of an escape opportunity. A chap in Electrical Design had apparently decided to leave Nissan – a rare occurrence – and had handed in his notice. So, they were short an engineer. Theoretically this was nothing to do with me – I already had an assignment and any changes should be negotiated through HR. But I was young and impatient, so I decided to bend the rules and go to see the Manager of the department – Andy Palmer. Variously called 'The Cheese' (a juvenile pun on the Japanese version of his name – Palmer-san or 'parmesan'), or simply 'AP,' Andy was already known as one to watch. He was barely 30, was already a Manager, and would soon be promoted to be the youngest General Manager in Nissan history. His friendly blue eyes, informal manner and easy grin could not fully disguise a fairly obvious sense of purpose in all that he said and did. I explained my problem, half-expecting him to tick me off, tell me to be patient and to follow the correct procedures in future. Little did I know AP. He heard me out, then replied with "Well, we can't have you buggering around in Trial Production, can we? I'll call your boss, speak to HR and it'll be sorted. You'll start with us in a week or two. Okay?" I was impressed – the world seemed very simple for this chap. He was as good as his word – I started to work in his department two weeks later, and would stay there for the next six years.

And so I started my first 'real' job – as a design engineer in the Electrical Design and Test department. During my time there I designed instrument packs, trip computers, switchgear, alternators and electrical sensors of various types. I was exposed for the first time to the world of the major automotive parts suppliers – companies like Bosch, Magneti Marelli, VDO, Siemens, Sagem, Yazaki. I found that I was very much at ease in the rather tense game of poker that car companies play with the supplier base. This relationship is fundamentally a dichotomy – the supplier and the car company have to establish a relationship, hopefully long-term. On the other hand, the car company is there to buy the part as cheaply as possible, and the supplier is there to sell it as expensively as possible, and to expend the least possible amount of expensive R&D manpower on it. Hence, there is inherent tension – and sometimes lots of it. I always looked on it as a game, and found I was good at it. Some engineers regarded the commercial side of things as 'Purchasing's job' and refused to get too involved in discussions about filthy Mammon. Not me. I enjoyed the often-fraught meetings with suppliers, the careful back-and-forth negotiation, the call-my-bluff poker-player psychology of it all.

The Electrical department, like the rest of NTEC at this stage, was very heavily influenced by Japanese expatriate staff. The original Nissan system of 'dual management' was still in place for a year or so after I joined – this meant that each local manager (Andy, for example) was flanked by a Japanese manager, who was there to give technical assistance, and to ensure smooth lines of communication with the mothership – Nissan Technical Center or NTC back in Japan. In the Electrical department our Japanese manager was Kataoka-san, an electronics expert whose only weakness was an unshakeable belief in putting in the long hours that were still very much a tradition in Japan. It was not unknown for him to actually sleep in the office, stretching out on the floor under his desk for a few hours before putting in another 18-hour day to solve some particularly sticky problem. Apart from the manager, each section, led by a 'local' Senior Engineer, had a Japanese *sokatsu* or 'senior technical advisor' attached to it. They were also there to provide technical support, especially in the application or interpretation of Nissan's in-house technical standards, some of which were still not available at this stage in English translation. They also provided a quick and efficient communication channel back to NTC.

It became clear to me later that Nissan had deliberately infused NTCE with Japanese industrial culture in a rather clever and subtle way. There was never any crass brainwashing, but there was steady drip-fed messaging of the right way to do things and the wrong way. The tradition of the Monday morning weekly meeting was strictly observed – 8.30 every Monday morning, each section standing up at their desks, listening to the Senior Engineer briefing us on the key news and activities at company level. Strange to Western eyes – a whole office-full of blue-uniformed people standing silently to semi-attention as the NCOs read out the orders – but to me it seemed normal, as I was exposed to it from my very first day. Nissan's corrupted version of the English language crept gradually and imperceptibly into our speech. 'Okay' was universally understood, but *not okay* was No Good, shortened to NG. The phrase *'NG desu'* or 'it's no good' in Japanese was used even in casual speech about a football match or last night's TV programme.

Other fundamental Nissan principles were taught directly on the job, often by direct instruction of junior engineers by the Japanese *sokatsus* and managers. One celebrated *sokatsu* – I shall call him Ogawa-san – was famous for his short temper with us green 'local' trainees. He did not hesitate to bawl us out like an old-fashioned sergeant major, puffing out his chest and scowling up at us from his five-foot-four vantage point. Ogawa-san's pet obsession was *genba* or 'checking with your own eyes': this meant, in essence, verifying facts personally, on the ground, and not relying on hearsay or third-party written reports. His message was – if there was a problem with your part on a car, get down there, find a car with the problem, and check it yourself – that is, with your own eyes. Then, if possible, check a few other examples to make sure that it was not an isolated incident. Sound common sense based on tried-and-tested principles of problem-solving, but delivered by Ogawa-san in a highly idiosyncratic manner. Almost all of us graduate engineers would at some stage cross his path with some problem or other, and be forced to report the concern. The following conversation would then inevitably take place:

Ogawa-san. "Hmmm. Did you check?"

Humble victim: "Yes, Ogawa-san."

Ogawa-san: "HOW did you check? With your own eyes?"

HV: "Yes, Ogawa-san," (although we all fantasised about it, nobody *ever* had the temerity to say 'No, Ogawa-san, I borrowed somebody else's eyes.')

Ogawa-san: "*So, desu* [is that so?]. HOW MANY did you check?"

At this stage, the game was over. If the victim said 'one' the answer would be, triumphantly "NOT enough! You have to check more!". The problem was that this was the guaranteed answer anyway, regardless of whether the victim claimed to have checked 5, 23 or thousands of examples – always using his own eyes, of course. Ogawa-san would always win, declaring triumphantly "NOT enough!" before dismissing the deflated victim with an imperious wave of his hand, back to the workshops to find more examples. Thankfully Ogawa-san was fairly unique.

I remember another part of my own engineering education, happily not at the hands of Ogawa-san but of our altogether gentler Manager, Kataoka-san. I had screwed up. The test results on one of the instrument packs that I had designed showed that the tachometer (rev-counter) calibration was wrong. Kataoka-san took me aside into a meeting room to gently quiz me as to how this had come about. Foolishly, I tried to blame someone else, explaining that the input signal data that I had got – from Engine Design section – was wrong, and that my part, the instrument panel processing unit actually did its job correctly. It was the input signal that was at fault. I did not quite use the phrase "It's not me, boss", but I came dangerously close, desperate to convince Kataoka-san that I really had done all that could reasonably be expected of me. Kataoka-san just looked at me, smiled almost sadly and said "Ah, but Dave-san, please remember that you are the designer of this part. You are responsible for *everything*. But *mondo nai* [no problem], you will know the next time" and patted me kindly on the shoulder. I quit the room red with shame. I was devastated – I would have much preferred an Ogawa-style chewing out than this mildest of remonstrations from this person for whom I had so much respect. At the time I was doubly frustrated because I thought that Kataoka-san was simply wrong – *how* could I be responsible for everything? It was clearly impossible – ridiculous!

Later I realised that Kataoka-san was giving me a very powerful message in his own low-key way: that as the design engineer *I* was the most important person in the creative chain – not my supervisor, not my *sokatsu*, not he as my manager, but ME, the designer. I had responsibility, and therefore a great deal of power – a very positive message of delegation and empowerment. Also, my responsibility was in fact much wider than my formal 'paper' responsibility for one part or one component. It was up to me – and nobody else – to understand the parts with which it interfaced, and how the whole system worked together. Taken to its logical limit, every part of a car is connected to every other part – it is one giant system. The engineer who has the intellectual curiosity and drive to understand 'everything' will eventually come to understand not only components but systems, and, finally, not only systems but the whole car. These two messages – which I retain to this day – were encapsulated in Kataoka-san's five words: "You are responsible for everything". Again, a message I only came to truly understand years later.

These four years spent as a basic design engineer were, genuinely, and without any rosy hindsight, the most enjoyable of my career. I was designing parts, and seeing them coming into production. I was working with many of the leading European and Japanese suppliers. I got the chance to travel for the first time in my life – mainly to Spain and Germany, where I, sometimes alone, represented the company in discussions that were worth many millions of pounds. I was young and had few commitments – my modest Nissan engineer's salary seemed like riches. I could buy a car and put a deposit on my first house. I was a member of a tightly-knit team in a company that was still small enough that I knew everybody by their first names. I was a boy scout – super-enthusiastic and always positive. And my assessments showed it – I was promoted quickly from graduate trainee to full 'engineer' status. I had already been marked down by Nissan's management as a quick learner and a potential future manager. But I was genuinely blind to this – I just wanted to design cars, or parts of cars, and that was enough for me. Even today I get a thrill out of seeing an old Primera or Micra built in these years parked in the street. I still cannot help crossing the street to glance at the instrument pack to make sure that it is still working okay – trip/total indicators still bright, dials not faded, the needles I designed all those years ago still properly aligned. And I still get a small thrill of pride when I see that yes, *my* bit still works, even if the rest of the car has usually seen better days.

CHAPTER 2

Apprenticeship

In 1998, six years after joining Nissan, I was promoted to Senior Engineer of the Test section of the Electrical Department. This was my first management job, in the sense of having people working directly for me. It was very much an 'in at the deep end' experience – the Test section was the largest in our department, with six engineers and eight or so technicians at any one time. Truth be told, I was not ready for the job – I was still too young and had no idea of how to manage people. I learnt human psychology and the basics of management the hard way – for both my unfortunate team members and myself – by experience. But there were other benefits, apart from starting to learn the imprecise art of managing people. I was now responsible not just for designing one part, but for validating the entire car, at least from an electrical and electronic point of view. It was a busy period – we continued to work on various versions of Micra, Primera and Terrano II, but we were also deep in the development of two new vehicles – the Almera, intended to be a VW Golf and Ford Focus rival, and the Almera Tino, a people-carrying variant, inspired largely by the success of the Renault Mégane Scenic 'people-carrier' version of the Mégane hatchback. So I had quite a wide portfolio of vehicles, and a lot to learn technically as well as managerially.

Shortly after I had been promoted, the management structure changed. The 'dual management' principle was abandoned – from now on there would be only one manager for each department, sometimes 'local,' sometimes Japanese. Kataoka-san returned to Japan – much regretted by his team. Our new boss was Koichi Suzuki, a quiet, unfailingly polite and unassuming professional. Six months after his arrival, we reached the 'job handover' stage of the Almera project. This was the point at which NTC in Japan would formally hand over the project to NTCE. The so-called 'production release' or final design approval of the production version of the components had already been done, and now Japan would formally delegate the final stages of production preparation to its European satellite. Suzuki-san planned to travel to Japan for two weeks to negotiate this job handover, and I was rather surprised when he asked me to accompany him – not only for the handover of my direct testing responsibilities, but also to negotiate the handover of the entire Department's activities. I was the most junior of his Senior Engineers, and I spoke little or no Japanese. It would have been more logical to have taken a more experienced 'local,' or one of his *sokatsus*. But I accepted his invitation gratefully without asking too many questions – this would be my first-ever trip outside Europe, and a not-to-be-refused first opportunity to visit Japan.

That first trip – in 1998 – was an eye-opener in many ways. We flew with ANA (All Nissan Airways) from London's Heathrow to Tokyo's Narita International airport – a journey that I was to make dozens of times in the years to come. It was my first long-haul flight, and I enjoyed it immensely, fascinated by the endless prairie of the Siberian tundra rolling by underneath the gently flexing wing of the Boeing 747. I followed Suzuki-san in a jet-lagged haze through the spotlessly-clean corridors of Narita Airport, to find our seats on the Narita Express train that whisked us rapidly and silently into Tokyo's immense western train hub – Shinjuku station. Shinjuku was the inspiration for some of Ridley Scott's *Bladerunner* film sets – a teeming, bustling chaos of huge skyscrapers, more people than you can imagine, flashing neon and blaring electronic sounds overlaid with apparently non-stop announcements in Japanese. One of the biggest train stations in the world, it can be overwhelming – especially for a first-time visitor, tired and bleary after a 12-hour flight.

Suzuki-san and I shouldered our way politely through the crowds. I felt as if I stuck out like a sore thumb – a too-tall, too-blonde tourist gazing uncertainly around me. I was reminded of the real meaning of the word *gaikoku-jin*, sometimes shortened to its less polite version of *gaijin*, and often politely translated as 'foreigner', but more accurately as 'outsider' or 'alien.' I truly did feel like an alien – that I was, for the first time in my life, in a truly foreign place. It was disconcerting, but very exciting at the same time. I drank in my surroundings, enthralled by every detail.

The next two weeks passed in fast-forward. It was, like everyone's first trip to Japan, a fascinating and exhausting experience. I am convinced that there are only two types of reaction to Japan by Westerners visiting for the first time – for some, the sense of dislocation and 'difference' is too much, and they fall into a defensiveness that borders on nervousness about the country. For others, it is truly liberating. I loved the 'otherness' of Japan from the first day, and that first business trip installed powerful memories that still make it a pleasure to go back there.

The Almera project handover exercise itself went very well – in fact it was largely ceremonial. I found that my very limited Japanese language skills were not as much as a handicap as I thought – most of the Nissan engineers spoke some English, and in any case very few common words were needed to communicate on purely technical subjects. In between the engineering meetings, Suzuki-san took me on a round of informal introductions, trawling around the desks of various Managers, General Managers and Directors in NTC. At the time I thought that this was a terrible waste of time – Suzuki-san had, surely, better things to do than take me – almost literally – by the hand to introduce me to these equally busy Japanese gentlemen. Only later did I realise that I was being groomed – in the best sense of that word. I had been taken to Japan partly to help Suzuki-san, but mainly to expose me to these executives – who would have an important influence on my future. One of Suzuki-san's jobs was to show them this young foreign 'high potential' engineer that NTCE claimed to have discovered, and to see if his face fitted. I tried my best to be presentable and trotted out repeatedly my few memorised stock phrases of Japanese *jiko-shokai* or 'self-introduction.'

Those two weeks were packed with the typical 'first-timer in Japan' experiences. At the weekend, my hosts kindly took me on some of the classic local tourist trips. On week nights, I was given terrifying passenger rides by my car-mad colleagues in various highly-tuned Nissan sports cars like R32 Skyline GTRs and S12 Silvias. I was of course invited to various after-work *nomikais* (literally 'drinking party' but actually meaning an extended dinner-party), accompanied by copious amounts of alcohol, during which the locals amused themselves by inviting me to eat the various delicacies that are usually trotted out to make foreigners squirm, and to test their mettle. I learned at how important these after-work parties were to the Japanese way of working, acting as an important safety-valve to the highly pressured and hierarchical environment in the office, bonding the work teams together in a haze of *sake* and shared laughter. I left Japan, having successfully concluded the handover 'negotiations' with Suzuki-san, with great memories and quite a few new friends.

Back at base in NTCE, work continued, seemingly as normal. The production lines in Sunderland and Barcelona rolled smoothly, sales figures seemed positive, and one rarely heard bad news. At working level, we genuinely did not know that in fact Nissan was in deep trouble. We were bleeding money. Lots of money – we would soon find out just how much.

The bombshell hit in the half-year financial announcements of September 1998. For the first time, the figures were bad. Very bad. We were in debt to the tune of 4.3 trillion yen or $35bn – telephone-number figures that were not in fact new, but had never really been communicated internally before. The international capital markets and investment banks, who had been prepared up until now to feed our debts, had finally got cold feet. Result – a cash crisis. And this in a period where the market was in its worst downturn for many years. The Japan domestic market in particular fell dramatically in 1997 and early 1998. The US market was still relatively strong, but even there, we were stumbling. As for Nissan Europe – well, we were the real bad boys. Nissan Europe had apparently been loss-making for every year since 1992 (the year I joined the company – coincidentally, I hoped!). I clearly remember the all-employee briefing that we had in late 1998 to explain this to us – charts bleeding red ink. We looked at each other stunned, nobody daring to ask the question that was on all our minds – how the hell did *that* happen?

For the next six months – until March 1999 – seemingly every week brought a fresh set of terrifying numbers. In November 1998 our credit rating was downgraded officially to 'Ba1' – a 'junk bond' rating. The Press was full of speculation on how one of the car world's giants could be brought to its knees, apparently so quickly. The collapse of the economy in Japan, and a slowdown in the USA was of course part of it. Nissan's obsession with blindly following Toyota had led to an explosion of product lines, platforms and powertrains. Unimaginative styling was also partly to blame for our sluggish sales, especially in Europe. Our cost structure was uncompetitive, not helped by our over-reliance on the *keiretsu* – the traditional Japanese network of supplier-partners, financially linked to the carmakers by complex cross-shareholdings. Our traditional, rigid Japanese upper management structure had not allowed us to react quickly enough when times got bad. Finally, the banks had been too quick to extend easy credit when the cash started to run short, allowing us to ignore the problem for far too long. But now the party was over – and we were left holding the bill.

In short, we were financially dead. There seemed to be only two ways out – either we would be nationalised in some way by the Japanese Government, or we would be taken over by a major competitor. The press was full of speculation about this. It slowly became clear that the Japanese Government had enough of its own problems, without having to turn its hand to car-making. But there were persistent rumours about a possible takeover by various other car companies. Talks with Mercedes-Benz were the subject of the most consistent rumours. Through the end of 1998 and into early 1999 the speculation bubbled on as we continued to work on in an increasingly uncomfortable limbo.

Against this rather depressing and very confusing background, I made another trip to Japan, in the spring of 1999. Once again, I was Suzuki-san's sidekick for a 'job transfer' from NTC – this time for the Almera Tino project, the MPV[3] or minivan version of the Almera hatchback. I found an NTC that had dramatically changed compared to my first visit. The cash shortage had hit, hard. There was no paper in the printers or photocopiers. Even the electricity was rationed as Nissan struggled to pay its bills. The lights and air-conditioning were turned off at lunchtime, and cut off again at 6pm, the official end of work. It was March, so not yet too difficult to live without the air-con, but the lack of lights was rather bizarre. Of course nobody left the office at 6pm, so when night fell, staff lit up some battery-powered desk lamps that they had brought from home and carried on working, the office eerily half-lit by the glow from the computer screens. Yet there was some glimmer of hope in the almost-literal darkness. One day, between meetings, one of the engineers took me to see the new EMC (Electro Magnetic Compatibility) chamber that was under construction. He took me on a tour of the facility – an enormous

---

[3] *Multi Purpose Vehicle, sometimes referred to as 'people-carriers' or as minivans in the US. Five or seven-seater vehicles with high roof-lines and typically with modular seats, designed as practical, safe, no-nonsense family transport.*

cavern, large enough to take the largest of trucks, where vehicles could be driven by remote-control on four-wheel rolling roads, while measuring their radio-frequency emissions, or alternately while bombarding them with extremely powerful electromagnetic 'attack' signals. It was a state-of-the-art facility, and obviously a huge investment. I asked my guide what it cost. He didn't know, but bent down to pick up one of the six-inch square, black, carbon-impregnated tiles that were used to line the chamber and hence seal it from all external radio-frequency emissions. "I don't know total, Dave-san, but each of these costs a thousand US dollars". I looked around me at the half-finished walls. There were thousands of these little black tiles already in place, and many more to go. I returned bemused to my desk in the sweltering half-darkness. We were in deep trouble, clearly, but this ambitious project gave me hope. Any company that would deny its own staff even light and cool air, yet continue to invest enormous sums of money in a long-term test facility, really, really wanted to survive. The stoicism of my colleagues, the almost-complete absence of complaint, demonstrated a commitment to the cause of Nissan that was almost frightening in its dedication. Not for the first time, I thought of Nissan as an army – we were on the defensive, almost over-run by the enemy, but we were not beaten. My colleagues would never, ever give up, with or without paper or electricity. It was strangely heartening. I felt even more proud of this strange company of which I was a part, battered and forlorn as it then was.

The press continued its feverish speculation about a possible takeover. The discussions with Mercedes-Benz seemed to be advancing. The occasional, guarded, press release seemed to confirm that the discussions between the Japanese and German big bosses were progressing well. I chatted about the situation with my Japanese colleagues. They were not happy that Nissan was about to lose its independence, but if we *had* to be bought by somebody, they could at least accept that it be a prestigious company with a rich heritage like Mercedes-Benz – they were, after all, the very company that had invented the automobile! We looked forward to the signature of the deal with something like resigned optimism. It was therefore quite a shock when we heard that all talks with M-B were off, and that Nissan's then President Hanawa-san was conducting 'very serious' discussions with someone called Louis Schweitzer, the head of Renault. More than one Japanese colleague approached my desk to ask me "Dave-san, what *is* 'Runo'. Have you heard of it?". I explained that yes, I had heard of it, that *Runo* was France's leading carmaker and a major player in Europe.

Renault was then almost unknown in Japan, its products sold in tiny numbers to rare Europhiles. They simply did not exist on the Japanese map of the automotive world. After two weeks, I concluded my business in Japan and flew home to learn that the Renault-Nissan Alliance had been formally signed. It was 27th March, 1999. Everything was about to change. Everything.

Many books and business studies have been written about the Renault–Nissan Alliance, or simply 'The Alliance' as it was to become known in-house, so I will not go into too much detail here. Suffice it to say that it is to this day a unique structure, neither a merger nor a joint venture, each company preserving its own distinct corporate culture and brand identity, but linking their destinies through cross-shareholdings.

Renault had launched a series of very profitable products through the 90s – firstly the Espace I and II, then the Mégane family, notably the Mégane Scenic MPV. As a result, by 1999, Renault had a very healthy bank balance. To set against that, it was still a very *Franco-Française*[4] company – strong on its traditional home turf of France, and in Europe in general, but weak internationally. The then Renault President, Louis Schweitzer, was convinced that survival into the 21st century meant becoming a true global player with presence in all global markets, and with the critical mass to resist local market storms. Hence his brainchild – the Alliance with an ailing Nissan. A daring move, especially in light of Renault's recent and painfully failed attempt to establish a partnership with Volvo. But Mr Schweitzer managed to persuade the French state (a 15% shareholder in Renault) as well as the sceptics on his own Board, and the deal with Nissan was signed. Press opinion was mixed – a minority of positive reports hailing the courage of Hanawa and Schweitzer, but most critics – fairly logically, it must be said – thought that Renault had gone crazy and would lose its financial shirt.

Between March and October 1999, we held our breath and tried to concentrate on our work in NTCE. But everything was in a state of suspense – we knew that profound changes were afoot. A new management team had been put in place to oversee the restructuring of the company. A 'hit squad' of approximately 30 Renault senior executives had been placed in key positions in Nissan Japan almost immediately after the Alliance was signed. This team was led by two big personalities: the initially unpronounceable Carlos Ghosn (Goz-N was a popular early mispronunciation, but in fact it's closer to 'Ghone') – an ex-Michelin business wizard with a reputation for stringent cost control – was named COO or Chief Operating Officer. Ominously, the Press reported (or fabricated) the nick-name of 'Le Cost Killer.' And Patrick Pélata, ex-head of Renault vehicle development, was appointed head of Product and Corporate Planning. Right from the start it was clear that these guys were not here to be passive internal consultants or advisors. Ghosn was here to restructure the overall business; Pélata was the product guru, brought in to revise our vehicle line-up. From the outset, two Golden Rules were clear. First, the two brands would be kept totally separate. To the customer, Nissan would remain Nissan, and Renault would remain Renault. Second, company culture would be respected – Renault had no intention of imposing a 'Renault Way' on

---

[4] *Franco-Française is a gently self-derogatory term, literally meaning 'French-Frenchy' used by the French themselves to denote a certain type of inward-looking parochialism.*

Nissan. Ghosn-san (as he immediately became known) and his team wanted to listen to the Nissan teams' own diagnoses of the problems, and would only impose solutions that were compatible with Nissan's corporate culture. This willingness to listen gave a very positive initial impression. Ghosn and Pélata seemed honest, open and ready to take action. But the analyses that were produced and communicated clearly to all staff (another Ghosn hallmark) were nevertheless very worrying.

Through now crystal-clear internal communications, we learned about the true profitability situation. Almost incredibly, Nissan had made the fundamental error of tracking the performance of each model on its Marginal Operating Profit or MOP – the 'sticker price' of the car, minus the direct cost to build it. This is almost always positive. But we had somehow forgotten, or rather, chosen to ignore, the Consolidated Operating Profit or COP – the profit based on the true cost of making the car – including the amortised R&D costs, the depreciation of assets like the factories and buildings, logistics and warranty costs, etc. Calculated like this, almost our entire range was loss-making. We were effectively giving away money with each car. This was deeply shocking – we all asked ourselves how we could have been so stupid. The reckoning for each region was bleak. The USA was barely profit making. Japan was slightly loss-making. But Europe was *deeply* loss-making, and losing market share in a plunging market. It was a money pit. Everybody in Nissan Europe – especially in NTCE – was deeply worried. It seemed obvious that intelligent businessmen like 'Le Cost Killer' and his lieutenants would do the only reasonable thing – shut Nissan Europe down and leave the European market to Renault. The dream of the early nineties that had attracted me to the company in the first place – to make cars 'designed by Europeans, built by Europeans, for Europeans' seemed dead, and almost naïve in hindsight. We were just waiting for the axe to fall.

The Day of Reckoning was 18th October 1999. Known internally at Nissan as 'Ghosn Day', it was awaited with great anticipation and trepidation. Ghosn's analysis of the situation and his recipe for restructuring was announced to all staff worldwide at the same time by video conference, broadcast live into every office and workshop in the Nissan empire. I gathered around the TV in one of our meeting rooms in Cranfield with a hundred others, all of us looking glum. Most of us were convinced that we would be looking for new employment after today. Ghosn took the podium and in his soon-to-be-familiar deceptively simple and direct way, outlined the now-famous Nissan Revival Plan or NRP, much-studied subsequently in business schools around the globe. Lots of logical actions – drastically reducing the numbers of platforms and powertrains, combining purchasing organisations and hence power, management restructuring. Then the tough part – plant shutdowns and job losses. Ghosn calmly and factually listed five plants for closure in Japan, with the associated loss of no less than 26,000 jobs. This was a thunderbolt. We knew it would be bad, but this was truly a cataclysm for Nissan in Japan. He moved on to operations in the USA. Here, still the most profitable region for Nissan, changes were mainly in company organisation. Next, Nissan Europe. We waited nervously. If Ghosn had wielded the axe so thoroughly in Japan, surely he would not – indeed he *could* not – hesitate to make deep cuts in Europe. But no – besides some technical reorganisation of the Nissan Europe management structure, and emphasising that purchasing synergies with Renault were doubly important in this region, he said nothing. No closure of plants. No mention at all of NTCE. Maybe he had forgotten – maybe this little outpost of 400 or so engineers was simply too small to be worthy of mention, and we were to be closed down after all? We checked – quickly – with our bosses. No, there was no mistake or omission: we were still in business. The relief was enormous.

Ghosn was very clear on another issue, however – from now on, no more loss-making products. This was translated into two very simple financial rules. First, all new cars would have to show a positive COP – usually 6-8% minimum. Second, *all* decisions would have to be profitable. This was checked by the application of a simple financial tool – the Nett Present Value or NPV calculation. A very basic tool taught in first-year finance courses, this compares the output of an investment to the return if the money were simply placed in the bank at standard interest rates. If the NPV is positive, it means that the investment is worthwhile. If it is negative, it's better to simply keep the money in the bank. From now on, *all* decisions in Nissan – from the purchase of a new office printer for 500€, to the investment in a new vehicle model for 500M€ – would be judged by NPV. These two simple checks had a radical, almost overnight effect on Nissan engineering. Previously, we had been technical purists, with a borderline-arrogant ignorance of financial calculations. Now it was up to us to apply some basic checks, using COP and NPV as our tools, and to take responsibility for the business impact of our actions. A small minority of engineers resented this. They were not capable – or willing – to see our job as anything other than a purely intellectual and technical exercise. But most people – including myself – took to this enthusiastically. The mathematics were simple, and the power of these new financial tools was obvious. It is not much of an exaggeration to say that Nissan was turned around by the systematic application of these two simple mathematical tools at all levels of the company.

By this time Andy Palmer, my old boss in Electrical, had risen through the ranks to become General Manager, replacing the Shingyoji-san who had so terrified me as a greenhorn in the Chassis department. Andy took to the new Ghosn regime like a duck to water. He had always been financially aware, a businessman as well as an engineer – now he took up Ghosn's message and insisted that NTCE become a real force in the fight back to profitability. Our first job was a simple, if difficult and unsavoury one. We had to cut parts costs – fast. This is the oldest job in the car world, and still one of the hardest. The question was brutally simple – without the time to redesign the model range completely, how to chop 5, 10 or 15 per cent out of the production cost of a car? Answer – enormous and aggressive pressure on the whole manufacturing chain – from raw materials suppliers, Tier 2 suppliers[5], the Tier 1 direct suppliers, and the production plant itself. A dirty job. A new man was flown in from Japan to head up this cost reduction drive, which was baptised '3-3-3'[6]. Akihiro Otomo was appointed Deputy General Manager, reporting to Andy. Otomo-san would be my boss for the next few years. Slight even by Japanese standards, he had a huge ear-to-ear smile and a ready laugh, which disguised an iron will and a very good poker face. Now, he had the thankless task of leading an NTCE team to cut swathes out of our cost base, fast. To do this he needed a deputy, a local manager. Despite my inexperience, I was appointed acting Manager, reporting to Otomo-san. This was, in fact, another 'Ghosn effect.' Pre-Alliance, Nissan had applied strict age-seniority rules for promotions. A Manager grade could only be assigned to someone who was in the age range 40-50, for example. These rules had been almost immediately scrapped ... from now on, promotion was to be on merit only.

To say I was not keen on the job would be an understatement. I really wanted to continue in direct design engineering work, and had been hoping to get a Manager position in one of the other technical departments, in order to learn more about the vehicle than its electrical and electronic systems. I also suspected – correctly – that this 3-3-3 task would be a bloodbath. But I liked Otomo-san, wanted to help him, and Andy gently twisted my arm as only he knew how to.

Otomo-san and I hand-picked a team of 3-3-3 engineers who were trained in financial tools like NPV, and we started to co-ordinate the cost reduction activities that were the Number 1 Priority of that first crucial year of the Nissan Revival Plan. Some of the work was interesting – for example, we had the opportunity to work with Renault in France directly for the first time, benchmarking the design and costs of various components, comparing Renault's future Laguna II (X74) with the final version of our Primera (P12). This revealed huge cost gaps – sometimes due to technical differences, more often due to Nissan's weak purchasing performance in Europe. Simply knowing that there was a technically unjustifiable gap often allowed us to resolve the problem in negotiation with the suppliers. We were no longer a tiny player in the European market – we were a partner with Renault, a very big player,

---

[5] *Tier 2 suppliers are the 'suppliers of the suppliers' – the companies that feed the Tier 1 suppliers like Bosch, Continental, Magna and Denso that sell their products direct to the carmakers. Tier 3 supplies Tier 2 ... and so on, and on.*
[6] *The 3-3-3 crash cost-reduction programme was so-named for the 'Three partners, Three regions, Three years' slogan. The partners were R&D, Purchasing and Manufacturing. The regions were Japan, North America and Europe. And three years was the amount of time Ghosn allocated to cut costs by 20%. Remarkably, that goal would be achieved in only two years.*

and together we were now the fourth largest car-maker in the world. For the first time, the big European suppliers started to listen when we requested something.

Other work was simply hard graft – the minutiae of generating thousands of technical cost reduction ideas, trawling through them to find the 10% of genuinely good ideas, then organising endless reviews to implement the design changes, carry out the testing, production trials and final application. Car companies are large, and almost by definition have a lot of inertia. They are set up to introduce big changes and new models efficiently. They are set up to *resist* small changes, however. Car plants are at their most efficient when they are stable – pumping out precisely the same product, over and over again. Each change is a risk to that stability – potentially introducing quality defects, or reducing plant efficiency (thus increasing manufacturing costs), or both. Hence, introducing thousands of minor modifications, many unrelated to each other, over an extended timescale of six or eight months, without the unifying drive of a new model launch, is very hard indeed. You quickly find yourself fighting the momentum of the company. This is what we did for this crucial year – a great deal of 'street fighting.' Much of the work was not particularly technically difficult, but was complicated, both politically and organisationally. I became Otomo-san's axe-man – the guy who delivered bad news, internally as well as externally, to the various suppliers. I made no friends in this job, but it toughened me up and taught me a lot about finance and cost-control; knowledge that was to serve me very well over the next years.

Andy was again as good as his word. Having paid my dues in the 3-3-3 team and helped the company weather the year 2000, the hardest year of its existence since 1945, I was duly rewarded with the next manager position that came up – in the Body Equipment Department. This was not my first preference – I would have preferred being boss of Chassis, or even Electrical, but I made the best of it. The Body Equipment team designed all the parts on the exterior of the car, except for the actual sheet metal itself. That meant lighting systems, wipers, external plastics like bumpers, spoilers and trim parts, as well as some odds and ends like badges and emblems. An eclectic mix of unglamorous bits, but an interestingly mixed bag in terms of technology. Huge cost pressure, as – with the possible exception of lighting systems – these are not parts that the customer notices, and hence values. They therefore have to be as cheap as possible, while being utterly reliable. For the first time I also had intensive involvement with Industrial Design or what was still then sometimes called 'Styling' – the folks that decide how vehicles and components should look, aesthetically speaking. The interaction between industrial designers and engineers is absolutely key to the success of a vehicle. My time in Body Equipment would start to teach me how that interaction works. I dived into the job with a will ...

CHAPTER 3

"Irasshaimase!"[7]

As the New Year of 2002 came and went, I was very much enjoying my job as Manager of Body Equipment. Granted, it had not been my first choice, but it *was* a 'real' engineering department, I had the coveted Manager title, and I had a respectably sizeable team working for me. I was doing okay in my career at Nissan so far – I had escaped the hard grind of the 3-3-3 team and the relentless drive for cost reduction, and liked and respected my bosses, Andy and Otomo-san. My staff mostly seemed okay with my being *their* boss – even the ones that were older than me and probably felt that they should have had a shot at the Manager role. The work was interesting – we were deep in the development of the second generation of Micra (K12) to be built in Sunderland, and I was very hands-on with that vehicle. As well as the new Micra, Nissan was still building the Almera and Primera in Sunderland, and the Almera Tino in Barcelona, and there was a constant flow of work to maintain these vehicles in production. I was an electrical engineer, not a mechanical engineer, so I had a lot to learn about the injection moulding of plastics, about boringly-functional-but-oh-so-important wiper systems, and about the precision engineering required to develop modern lighting systems – headlights and rear lights.

Meanwhile, Nissan was still in bad shape, especially Nissan in Europe. Granted, the emergency defibrillation of the company by Carlos Ghosn, Patrick Pélata and their little cohort had worked – the hard graft of the Nissan Revival Plan between 1999 and 2001 had restored the company's heartbeat. We had indeed turned a net profit in 2000 – 331 billion yen (approx $3 billion) on global sales of 2.6 million vehicles[8] – the first profit for a decade. And by the end of 2001 things *were* looking up – profits up to 371 billion yen (approx $3.4 billion) on the same number of vehicles sold. Nissan Europe, however, lagged behind Nissan's 'domestic' sales in Japan, and far behind our overseas colleagues in North America. The latest 'P12' version of the Primera was dying in the market, buyers alienated by its radical styling

---

[7] *Irrashaimase means 'welcome' or 'come on in!' in Japanese. This is the word shouted by staff in Japan to welcome newcomers into a store or restaurant. It is one of the first words that all visitors to Japan will hear. As I moved to work for Nissan in Japan, even after years of working for Nissan in the UK, I felt very much like the new boy being invited into the back room of the company 'shop' for the first time.*
[8] *All figures from Nissan's 2000 and 2001 Company Reports, https://www.nissan-global.com/PDF/ar_fy00e.pdf.*

and by the lack of a competitive, modern diesel engine. It was losing out in the sales battle to stronger competitors like the BMW 3-series and the Mercedes C-Class, as the premium German marques moved 'downmarket' to dominate the medium-sized family car market as they already did the larger, more luxurious segments.

Almera sales looked reasonably strong on paper, apparently holding its own against its competitors like the VW Golf and Ford Focus, but unfortunately it was a loss-maker. The car was a profit black hole – of which more later. Almera Tino, the small people-mover version of the Almera that we built in Spain, simply could not compete against our Alliance partner's equivalent, the Renault Mégane Scenic too vanilla in flavour to persuade customers to ignore its cooler French cousin that had a certain *je ne sais quoi*. Thus, the only model we had in the line-up that was a real success was the little Micra ... so a lot depended on the new 'K12' model that we were working on. But 'small cars, small profits' says the old adage of the car industry ... and it's not far from the truth. So, Nissan Europe was feeling the pressure. We knew that Carlos Ghosn would not allow an entire region to continue bleeding money. Something had to be done ... and that something had to be a better, more popular, and vitally, more profitable model range.

March 2002. It was against this background, while I was on a business trip to France, that I was offered the opportunity to contribute very directly to this challenge. I was visiting Valeo, one of the world's largest car parts suppliers, and a specialist in lighting systems. One of the trickiest issues around automotive lighting is, believe it or not, fogging – or more precisely, *anti*-fogging: how to prevent that light misting effect that you may have seen around the edges of the lenses on the headlights of a car, usually older cars, and hopefully not on any of the cars I've been responsible for. The problem is a simple one, but surprisingly hard to solve. A headlight cannot really be fully sealed – the light-emitting element, be it a normal old-fashioned bulb, a halogen tube, a Xenon unit or even, these days, a powerful LED unit, generates heat. This makes the gas (air) inside the headlamp expand. If the unit were truly sealed, it would burst under this pressure, unless it were to be built of unnecessarily robust materials. The solution is to let the headlamp 'breathe' by building small vents into the housing to let the air escape. But, inevitably, the headlamp will cool at some stage – probably when you park your car after arriving at home at night. As you unlock your front door, your headlamp, its duties done, is slowly cooling down in the dark night air. Unfortunately, that dark night air is often also damp night air – especially in Northern climes. So now the little vents that were built in to let the air *out* are doing the reverse – they are letting humidity in as the gases in the lamp contract and 'suck in' the air from outside to fill the void. Sadly, as the air cools, it loses its capacity to hold water vapour – and that water is deposited on the inside of the headlamp lens, as the misting effect that we don't like to see – it makes our car look like an old dog: and nobody wants their pride and joy to look like a 12 year-old mutt. The solution? Various one-way valves and membranes, like the well-known Gore-Tex®, so that the vents allow dry, hot air to escape, but block the water vapour in the cold air entering when the headlight cools down. The position of the vents relative to the light-emitting heat source is also crucial – and carmakers and their specialist lighting suppliers like Valeo, Hella or Ichikoh spend enormous sums of money modelling the behaviour of the headlamp in various situations to get this right. In recent years this has become easier – highly-efficient LEDs emit more light and less heat, and hence the expansion and contraction of the air within the headlamp is less of an issue. But remember that we are back in 2002, and LEDs powerful enough to use for headlights were yet to be developed. Hence, I had travelled to Valeo's technical centre near Paris to review computer simulations and physical test data for various headlamp fogging issues on the new Micra, as well as some of our other vehicles already in production.

I was staying at a modest hotel in Versailles, to the south-west of Paris, as I had some meetings at Renault's giant *Technocentre* R&D base the next day. It was a typical Nissan-losing-money-days era hotel – 2 stars if we were lucky, no frills like restaurants or anything fancy like that. We had dinner with the Valeo folks, always proud to show off the local cuisine to visitors. The day had gone well, we had had some good discussions, and I was comfortably full of good Parisian food as I walked into reception of the hotel around 11pm. I picked up my room key and the receptionist said "Mr Twohig, someone left a message for you," in English much better than my then fairly non-existent French. She passed me a hand-written bit of paper which simply said 'Call Andy' with his telephone number. The word 'Urgent' was tacked on as an afterthought. I thought it over as I made my way to my little bare room – it was pretty late, maybe I should leave it until tomorrow? But then that word 'urgent' swayed me – Andy would be pretty annoyed if I did *not* call, if it really were urgent. So, I made the 'phone call that would change my life.

Andy picked up after a few rings: "Ah, Dave – thanks for calling me back."

"No problem, Boss. What can I do for you?"

"Great news. We've got it. Case 3 project. NTC have only bloody given us the lead on the Almera replacement!"

Now, I can't overstress how exciting this news was. But first, let me explain what this 'Case 3' jargon meant. When Nissan set up its overseas R&D operations in the late '80s, it had no intention of simply letting a bunch of foreign engineers loose to design its cars from the ground up, with little or no control from the mothership. So, it invented three 'Cases' or scenarios for overseas development. The first, Case 1, meant simply copying parts strictly to Japanese designs with local suppliers – simple parts localisation. If that went okay, the overseas organisation would be allowed to make limited local versions, still of vehicles engineered in Japan – for example, to lead the development of a left-hand-drive version of a vehicle, or to install a diesel engine in a car that had been engineered in Japan for petrol-power only. The next step was being allowed to lead more significant variants of base cars – a good example was the estate or 'wagon' version of the Nissan Primera that NTCE had been allowed to develop a few years back. The next stage of delegation was Case 2, where the local R&D centre would be allowed to *complete* the development of a car that had been started in Japan. Typically, the basic engineering would be done in Nissan Technical Center (NTC) in Japan, then handed over to the overseas team at a certain stage – usually when the first prototype vehicles had already been built and the design was fundamentally proven. These Case 2 projects usually involved sending a Joint Development Team (or JDT) to Japan for a period of six months to a year, to take the project handover and bring the project home successfully. NTCE had in recent years carried out two Case 2 projects, and despatched two JDTs to Japan – the first working on the Almera, the second on the third-generation or 'P12' Primera launched in 2001, which was an unmitigated commercial disaster. But now, at last, Japan was offering us the Holy Grail – our very own Case 3 project. Case 3 meant that Nissan Japan would not only allow us to *build* a new car in the UK, but also to design and engineer the car ourselves. No 'handover' involved – it would be ours, from day one. This was Japan letting us off the leash – fully 14 years after Nissan had established the European Technical Centre in the UK. It was the culmination of a long, hard, slow apprenticeship – proving our worth as engineers the hard way, through the patient years of Case 1 and Case 2 projects.

I knew that Andy had campaigned hard to get us permission to 'do' a real Case 3 project. I also knew that there was a little friendly rivalry going on here with our US-based rivals – Nissan Technical Center – North America (NTC-NA) in Detroit: they were the only other 'overseas' Nissan engineering organisation, and we always felt slightly junior in comparison. As mentioned above, Nissan's sales – and profits – in the US dwarfed ours in Europe. NTC-NA had been established two years earlier (in 1986, compared to 1988 for NTCE) and most annoyingly of all for us Europeans, had already been granted its first Case 3 a year or so earlier, with the project that would become the highly-successful Altima, built in the giant Nissan plant in Smyrna, Tennessee. This piqued Andy no end – his competitive spirit wanted to persuade Nissan Japan to give *us* a shot at a Case 3 project – and now he was telling me that he had succeeded; that NTC had finally agreed to give us a shot at the title.

"So, we're going to be sending a team over there – no bloody Joint Development stuff this time, we're in the driving seat. This is THE development team. We're going to be running the show."

"Great!" said I, "How many lads will be going over?"

"20 or 30. Going to be a hell of a lot of work to do."

"You're not kidding. So where do I come in?"

"I'm going to need someone chippy enough to lead the rabble. You're one of our candidates … you're bolshy enough. Plus, you seem to get on okay with the Japanese."

I felt the excitement rising, despite myself. This would be a plum job – the real thing, at last. No messing around correcting the work of others. No wishing that the 'base vehicle' was different. *We* would be doing the base vehicle – if it was good, the credit would be ours; if it were a piece of junk, the blame would be ours. I already wanted this job, very badly. But I knew Andy well enough to know that he was probably testing me in some way. And I really did not want to fail his test.

"Sounds great. When does the project start?" I asked, cautiously.

"Ah, that's the thing. Right away. No messing around. We need the team to be in Japan in early April, just after they do the yearly organisation musical chairs thing."

This was an eye-opener. Every year, Nissan in Japan traditionally did a job rotation for many key posts on 1st April. This was a very logical date for a team to take up a new project. But it was barely a month away! How the hell were we to get 30 engineers signed up and shipped out to Japan in that space of time? I was sceptical, but knew better than to voice any hesitation. 'Can't be done' was not a phrase to roll out lightly with Andy.

"So, are you up for it?" continued the tinny voice on the telephone. "A year in Japan, heading the team. If you don't screw it up too badly, I'll make you a GM when you come back."

My suppressed excitement went up a notch. I was ambitious and competitive enough back then to know that the 'GM' or General Manager title was a big deal. Andy had been one of the first foreigners to reach that grade through internal promotion, when he was still in his early thirties. This was unbelievably young by the old Nissan standards – in the pre-Ghosn days, GMs were older, Japanese gentlemen with veins of frost already running through their hair. The blond-haired, blue-eyed Andy was a phenomenon, achieving this elevated rank in his early thirties. Here he was very clearly goading me – dangling the carrot. And I was young and foolish enough to reach for that carrot …

"Sounds good" I replied." Let me think it over. Can I call you back first thing tomorrow?"

The next thing he said made my blood run slightly cold:

"Yeah, think it over. You have 10 minutes to call me back. Then I call the next number on my list."

His voice was deadpan. I knew from his tone that he was not joking. He was deadly serious. Another test. I was annoyed. But I knew I had no choice.

"Okay. I'll call you."

"Great. Ten minutes, yeah? Talk soon." Click.

I paced up and down the poky room. I was in a serious quandary here. I was fuming at Andy. Why the hell did he have to put me under this pressure? This was a lose-lose situation. If I said no, well, Andy would call the next guy on the list – whoever that was. If I said yes, without checking with my wife, Cooleen, I would also be in deep water. I swore under my breath and above it, then stopped wearing a trench in the cheap hotel carpet. Seven minutes later, I re-dialled Andy's number.

"Well?" he said ... I swore I could almost detect a hint of a smile in his voice. He was loving this.

"Okay. Count me in."

"Good man. I knew you would. Better start packing when you get home." He was laughing openly now. "By the way, congratulations on the promotion."

"Eh? What do you mean?"

"Congratulations on the General Manager stripes."

I was confused as hell now.

"You mean, when I come back from Japan?"

"No. Don't be daft. I'm not going to send the team leader out there as a Manager. You need 'GM' on your business card for the Japanese guys to take you seriously. You still look like a snot-nosed kid. But I didn't want you to say 'yes' to the job just for the promotion. I wanted you to say yes for the right reason – because you wanted to do the car."

"Cheers, Andy. You know my wife will bloody kill me, don't you?"

More chuckling.

"You'll talk her around. Turn on some of that Irish charm."

"She *is* Irish, Andy. Irish charm does not work on her."

"Goodnight, Dave. Come see me when you get back."

"'Night, Boss. You bastard."

He was still chuckling as I put the phone down and tried to get some sleep. This was big ... to say that I was excited would be an understatement. And yes, my wife *did* kill me for deciding to take us both to Japan for a year without consulting her – she still does not see the funny side of this story.

Back at base in NTCE a few days later, it was quickly announced that I would be appointed General Manager and Assistant Chief Vehicle Engineer (or ACVE) for a new Case 3 full-vehicle project, code-named B32A. Bear with me – a quick explanation of the role of ACVE will be required here. All companies like their three- and four-letter acronyms, and Nissan was no exception. The CVE or Chief Vehicle Engineer role had been created soon after the Alliance – another part of the re-jigging of the whole organisation. Nissan folks tended to believe that it was another French import from Renault, but this was not in fact the case. It was a genuine Ghosn-era innovation.

CVEs were basically given full authority over all aspects of developing new vehicles. They were fully responsible for the so-called Iron Triangle of Cost, Quality and Delivery (CQD) so beloved of post-WW2 management theorists. They ran the whole engineering 'shop,' and it was their responsibility if the project ran over cost, did not reach production in time, or if there were any quality issues – even many years after the car was launched. The hundreds of engineers of various disciplines required to engineer a vehicle did not all report to the CVEs hierarchically – the teams were organised in the classic two-dimensional matrix organisation structure. The head of the Powertrain engineering team (or 'line function' in Nissan jargon) for example, would be the direct, hierarchical boss of all the powertrain engineers, working on many different vehicles. But part of his team would report on a 'project' basis to the CVE responsible for a given vehicle. But the CVEs were clearly the Big Bosses within Engineering – Ghosn had made it very clear that these guys (predictably, they were still all men back in the early 2000s) were in charge of things now, and the buck stopped with them. That gave them enormous clout – they had the blessing of Ghosn-san, and that was like a superpower within Nissan at the time And in fact that was how I heard the role of CVE first described – as an 'engineering Superman'. But the CVEs were not quite responsible for *everything*. They did not decide how the car looked – at least, not directly. That was the role of the Chief Designer, reporting to the head of the Design Studio. They also did not control the pricing or the market positioning of the vehicle – that was managed by the Product Planners, and very closely controlled by the Program Directors (PDs).

The Program Directors were ultimately responsible for the all-important project profitability. In short, the PD had the enormously difficult task of ensuring that the cash *in* (from sales – a simple function of the amount the customer was willing to pay for the car, multiplied by how many you could sell over its lifetime) exceeded the cash *out* (the amount of money spent by the CVE in designing and building the cars). So these two positions – CVE and PD – held enormous power and prestige. And of course, they were all in Japan. A CVE would typically manage all the cars on one 'platform' – and that could mean up to seven or eight vehicles. This optimisation of platform-sharing had become of capital importance, since the Nissan Revival Plan had slashed the number of platforms in half, from 24 to just 12. The CVE's job was to ensure that we could build as many cars as possible, with the highest possible commonality of engineering and components, on each of these remaining 12 platforms – the exact opposite of what Nissan had allowed itself to do in the bubble years of the 1980s and 1990s. The *Assistant* CVE or ACVE – my new job – was the CVE's right-hand man, responsible for *one* of these cars on a given platform. He was expected to live, breathe, sleep his project, ie his car, and he had full delegation for the success or failure of it. The CVE would help out, and would keep a close eye on the platform, especially any deviation from platform commonality with the other cars sitting on that platform, but the ACVEs were expected to be big boys[9], and to live or die (professionally speaking) on the success or failure of their vehicle.

Now, you may be wondering why the new Case 3 development team had to decamp to Japan in the first place. Why could we not have simply done the work in the UK? Well, there were two reasons for this. The first, official, reason was that the initial engineering work on all cars is intimately linked with platform and powertrain selection – and those platforms and powertrains were shared with other cars to be built and sold in Japan and the US. Therefore, the early work on any car had logically to be carried out in a central R&D location with a global purview – NTC. The second, less official but arguably more important reason, was to train us. None of us had ever designed an entire car from scratch before. Nissan believed in 'on the job' immersive training – learning by doing, but under careful supervision, by those who *knew* what they were doing. So for one year, we would work under the informal but close supervision of our Japanese counterparts – and that supervision applied very much to me also. I was a newbie at this game, and a lot was riding on it. I was therefore to be apprenticed to the best of the best.

My new mentor and boss was the *capo di tutti capi*[10] of all Chief Vehicle Engineers – Kimiyasu Nakamura. In fact, I now had *two* bosses – Nakamura-san (or Nakamura-CVE as he was commonly known – the -CVE honorific often replacing the standard -san internally in Nissan) and Andy, by the system that Nissan referred to as 'dotted-line' or double-reporting. I had never met him, but had heard a lot about Nakamura-san, or 'Kimiyasu' as only those closest to him could call him. He had actually worked in the UK with NTCE, back in the 'old days' of 1988 to 1991 – just before I joined the company. He had been one of the first *sokatsu* despatched to train the new European recruits, and was later appointed Manager of Body Design. Those who knew him and had worked for him spoke of him in almost-hushed tones of respect. He sounded like a formidably competent engineer and was also apparently an excellent people-manager – a rare combination. I was intimidated by his reputation before I ever met him. I was to realise later that all the rumours and the hushed tones of respect were justified. He was – and still is – the best manager I have ever encountered.

Almost immediately, I had engineers coming to me to volunteer for the team that was to go to Japan. I was not the only one who had been waiting for a Case 3 project – people knew that this was their big chance to shine. Most of the volunteers were young. NTCE in any case was a very young company – the average age at that time was about 30. I was already older

---

[9] *Please bear with my repeated use of the male pronoun, but in the early 2000s, Nissan was still a very male-dominated company. All of the PDs, CVEs and most of the senior managers were still male back then.*
[10] *'Boss of bosses' in Italian.*

than most, at 32. But the timeline meant that most of the volunteers were young and single. We had to be in Japan on 2nd April. That meant upping sticks in one month – nigh-impossible for married folks, especially those with kids. Selling or renting out houses, finding English-language schools in Japan, managing spouse's jobs in some cases – that was a near-impossible ask in just one month for those with families. Luckily, I had more volunteers than there were places on the team. I was therefore able to hand-pick my team to some extent. Of course, I did not have complete control over this – the NTCE 'local' managers got some say in who could go, and who had to stay behind. But generally, I was very happy with the team that slowly coalesced around me in the next few weeks, as the clock ticked down to 1st April and our take-off date. They were all hungry, ambitious in the right way, and – almost by definition – up for a challenge. Anyone who signs up, at a month's notice, to go to work in Japan for a year, on an unknown project and for a complete beginner (ie, me) needs to have a certain risk tolerance. And all of my team did. This was to prove to be a key asset – it turns out that sharp *focus*, a key aspect of developing any successful vehicle, automatically implies risk-acceptance. And as it would turn out, the next few years were not for the faint-hearted …

I was extremely fortunate that a few of the older, more experienced senior engineers *did* manage to do the impossible and arrange to move to Japan with their families. Some of these volunteers formed my immediate core team, helping me manage the less-experienced engineers, and any credit for the success of the projects we worked on is in large parts, theirs. Matthew Ewing was my rock. Matthew was a safety engineer and in many ways my opposite: calm, soft-spoken and thoughtful, he was the yin to my yang – reflective, measured, and steady, where I was, well, none of these things. Married with a couple of young kids, Matthew would be the people-manager – most of the 25-odd engineers would actually report to him directly, so that I would concentrate on pure project management. Peter Brown was our most experienced team member – again, a family man who had uprooted his kids and moved them halfway across the globe at a few weeks' notice. Peter was Manager for what was known in Nissan as 'Marketability' – not to be confused with Market*ing*. This is the engineering discipline sometimes known as Customer-Oriented-Engineering or COE – the science, craft and art of measuring customer needs and engineering the car to meet them. Very experienced, Peter had the rare gifts of 'feeling' what the customer needed, being tough enough to drive engineers to deliver those needs, but also having the emotional intelligence and humour required to avoid alienating those engineers. It's easy to be a demanding hard-ass: Peter had the rare gift of demanding excellence from the team members, while making them feel good about it. Everyone liked and respected him. The third member of my core team was Roger Blakey – an electrical engineer like me, we had already worked side-by-side for years. Roger was another family man who managed the impossible to come to Japan with me – and had signed up to be the Project Manager, helping me keep tabs on the various Quality, Cost and Delivery metrics. If I was Nakamura-CVE's right-hand-man, Roger was to become *my* right-hand-man. A blunt-spoken call-a-spade-a-bloody-shovel Yorkshireman, Roger became my enforcer – he policed the cost and timing of the project with a rod of iron, and would become the bane of any engineer wanting to add a cent to the cost of a component, or suggesting that he might need an extra day to hit a project deadline. Roger was a guard dog – but unlike most such 'enforcers' within the car industry, he knew how to temper the toughness with a sense of humour. He was (and still is) one of the funniest people I know, delivering killer lines in a flat northern drawl that took the edge off the tough job that he had to do.

These three men would be the core team that drove the success of the project that we would eventually deliver together – Matthew binding the team together, Peter making sure that we never forgot the end customer, and Roger keeping the discipline we needed. Knowing that these three 'had my back' allowed me to totally concentrate on the vehicle itself. I was absolutely overjoyed that Dave Kelly – yes, the same Dave Kelly that had so impressed me with his induction lecture ten years before – also managed to join the team, again having to subject his family to all the upheaval that meant. I was delighted to have this solid, experienced and deeply competent fellow countryman with me. At least I knew that I would have absolutely nothing to worry about in terms of Body design with Dave on the case. We'll come across some of the rest of the key team members later.

The last month before our departure passed in a whirlwind of working with HR to set up 'foreign service assignment' contracts, arranging working visas with the Japanese embassy in London, reassuring local Managers, making arrangements for IT facilities – computers and CAD 'tubes' – in Japan, and generally getting to know my new team as well as I could before we left. I had very little information or technical briefing on the actual project that we would be leading – this vehicle code-named B32A. I had just one telephone call with Nakamura-san, and he did not tell me much about the actual vehicle. His English was perfect, with a slight accent, detectable to me as a Japanese person who had spent time in the UK (we had pretty finely tuned ears, and could tell those who had been on assignment in the USA from those who had been in the UK by their accent and phrasing). Soft-spoken, very polite, he told me that he was really looking forward to meeting me and to not worry about the project – there would be plenty of time to brief my team and me when I arrived in Japan. He wished us safe travels and was gone.

The big day arrived before I knew it. I decided that I should fly out a few days before most of the team, to make sure that everything was in order.

My wife had arranged a sabbatical from her work in the UK, but she could only manage six months' leave – a year was too long for her company. We decided that I would go out alone for three months, then she would join me for six months, and I would spend the last 3 months 'solo' again to wrap things up in Japan. Hence, I found myself alone in Heathrow airport, a couple of big suitcases with me, waiting for my one-way flight to Japan. I had taken the trip enough times by now to make it familiar, but this time it had special significance, knowing that I would now be living in his country that so fascinated me, not just coming for a couple of weeks' business trip. So, I peered out of the window with more than my usual curiosity as the ANA 747 cruised past Mount Fuji on our port side, then turned northward, crossing the Japanese coast and headed inland to Narita Airport, undercarriage thumping down reassuringly as the characteristic blue-tiled roofs and rice paddies of the Japanese countryside flashed under the wings. I watched the politely busy traffic and unique Japan domestic-market-only vehicles – little square minivans, cute *kei*-cars[11] and the stuck-in-the 1970s taxi cabs – as my shuttle bus made its way across the incredible urban sprawl of Tokyo, a billion lights winking on as the sun set across this megapolis of 35 million souls. The bus dropped me off at Hon-Atsugi train station, and I trawled my unusually-large suitcases up Ichiban Street, between the usual pinging sound effects of the Pachinko parlours and the electronic welcome messages of various mobile phone stores to the standard 'Nissan' hotel, the Royal Park.

I had done this trip many times by now – so it was strange to think that this place was now 'home' and that I would not see Europe for a long time. I am lucky; I sleep pretty well on aircraft, so long flights do not hit me too hard. But I still managed to persuade myself that my inability to sleep that first night was down to jetlag. In truth, I was unbelievably excited to start the job, to finally get my teeth into this vehicle that my team was now fully responsible for – this mysterious project that was, so far, just that anonymous alphanumerical code: B32A.

---

[11] *Tiny, 660cc mini-cars unique to Japan, encouraged by specially favourable tax laws.*

CHAPTER 4

False start

It was early April, 2002. My team of 25-odd volunteers and I were now based in the rather ominously named Building 101, the main Engineering building at the heart of the Nissan Technical Centre, a huge engineering campus and the working 'home' of close to 10,000 engineers. It was set in the surprisingly leafy green hills of Hon-Atsugi, a small (by Japanese standards) town that had been absorbed into the very western edge of the Tokyo megapolis. Tokyo is an odd city – it does not peter out in an endless sprawl of suburbs and industrial estates, it just stops, dead, and the steep hills of the Japanese countryside spring up, green and pristine – humid and jungly in the summer, starkly beautiful in the winter under their mantle of snow. NTC was built in the early 1980s in a bowl set in these hills. The main access is through twin tunnels drilled through the rock of one of the surrounding hills, worthy of the best James Bond movie-villain lair, and making arrival at this 'secret' R&D site quite dramatic. Doubly so for the NTCE team in the spring of 2002, arriving here to make our mark, and to prove just what Nissan's European upstarts could achieve. This was to be our team's base for the next year or so, and was to become a professional home from home for me for the next few years.

I arrived bright and very early on my first day, 1st April 2002 – fortunately forgetting that it was April Fool's Day – to be met at visitor's reception by Nakamura-CVE's assistant, Katoh-san, who hung a garland of electronic security passes around my neck and ushered me through the labyrinthine building. We crammed into the lift and I remembered basic Japanese lift etiquette[12]: never take the lift for less than two floors; always stand near the control panel and operate the floor and open/close buttons yourself – leaving the rear corner, furthest away from the buttons, for more senior people. I had the usual polite struggle with Katoh-san, who shushed me gently but firmly away from the buttons, as I was theoretically more senior than her. Always mutter *sumimasen* (excuse me) as you get in. A little bop-of-the-head bow to the other passengers as you get out. Katoh-san took

---

[12] *There are innumerable such unwritten rules – not only how and where to stand in a lift, but where to sit in a meeting room, where to sit in a car (most junior person drives, of course) where to sit in a bar or restaurant, etc. As a foreigner you will make basic mistakes, but it's much appreciated if you make at least a fumbling effort to get it right*

me on a tour of the fifth floor of the building, where Engineering Project Management, of which I was now part, was based. A classic Japanese office layout of the time – a seemingly endless, rectangular space, hundreds of metres long, maybe 50m wide, with a central stairwell and a rank of lifts at each end. The desks were set out in incredibly closely spaced rows, at right angles to the walls. And in a row along the outer walls, with their backs to the windows, were the bosses' desks – the General Managers and above, who had earned the right, through decades of loyal service, to escape the tightly-packed parallel rows and work with their backs to the natural light, looking out over the row of desks occupied by their immediate team. Katoh-san led me to Nakamura-CVE's desk. He was not there yet, but I saw, to my embarrassment, that a small 'mini-boss' desk had been set up, literally a few inches from his desk, at his left elbow. It was an inch or two lower than his, and several inches narrower, but it was unmistakably the apprentice's desk. This was mine. As Assistant CVE or 'ACVE' in Nissan-speak, I was no longer Dave-san, I was now Dave-ACVE, and some of Nakamura-CVE's gold dust of power would hopefully rub off on me, by default. I installed myself at my little desk, feeling horribly self-conscious, and aware of the curious glances of the team members as they filed in one by one, the desks started to fill, and the office hummed into life.

I set off on a tour of the building to make sure that my team members were finding their places okay. Matthew, Peter and Roger, as my immediate team, were also based in the Project Management area with me. But the other members of the team were to be physically based with their respective technical families – the Electrical guys in the Electrical team area, the Body guys in the Body section, etc. This meant that the team was physically spread out over several floors of this huge building – a great advantage, as it turned out, allowing me to practise the proverbial 'management by walking about'; a valid excuse to wander the spaces of this engineering beehive and hence keep abreast of whatever was going on. Having confirmed that the guys were more-or-less okay and were settling into their local teams, I made my way back to meet my boss – the famous Kimiyasu Nakamura.

How can I describe him? Nakamura-san at first glance is not particularly remarkable. Of medium height, he's a typically well-groomed Japanese businessman – sober suit, white shirt, perfectly centred tie. The suit jacket was hung carefully on a hanger and replaced every morning with his light-green NTC version of the Nissan 'blues' uniform. His English, as I had already noted on the telephone, was perfect, polished by his years in the UK. He was politeness incarnate – in years of working with him, I never once heard him raise his voice or lose his patience, never mind his temper. He was physically very still – I can still picture him at the start of one of the many meetings we would attend together. He would sit, place his metal flask of *o-cha* (green tea) to his left, the small pen-and-pencil pouch he always carried to his right, place his fan[13] parallel to it, either fold his arms or pick up whatever papers might be on the desk, and then sit *still*. I mean still, still. I do not know if it was a natural habit, or if it had been deliberately cultivated, but it immediately conjured an air of calm authority in whatever room he entered. As soon as Nakamura-CVE arrived, set out his stall – almost literally – and said *"Hajimemashou ka?"* ('Shall we start?') one knew that we were in business – no more messing about. It was one of the many things about Nakamura-san that I tried, and largely failed, to copy. In many ways we were opposites, and as a fidgety, hand-waving not-particularly-calm Irishman, I was never going to pull off Nakamura-san's air of quiet, confident stillness.

I would learn in the weeks and months to come that his power and authority did not come from simple tricks like knowing how to sit still. He had – and probably still has – a phenomenal memory. I would regularly show him a layout drawing for a car, with dozens if not hundreds of dimensions on it – track, wheelbase, position of seats relative to floor, etc. He would look at it once, ask me a few questions – maybe on the mass of the car or some dimension in particular – and just nod as I gave him the responses. What was amazing – and quite intimidating was that he would remember these figures, to the millimetre or kilogram – months later, be able to recall them in meetings, or (worse) to correct me gently if I could not recall them myself. It was the same for cost figures – he could retain the costs of parts and whole vehicles to uncanny levels of accuracy. I have a pretty good memory myself, especially for costs, but Nakamura-san made me look like an amnesiac. And bear in mind that he was responsible for maybe 6 to 8 vehicles at any one time – my European project was merely one of them.

Nakamura-san had two other 'superpowers.' The Nissan technical staff deeply respected his technical judgement. This was absolutely key in an engineering community that was – despite the trauma of the collapse of Nissan and subsequent painful rebuilding – very proud of its technical chops. The Nissan engineers still considered – rightly or wrongly – that they were the equal of any engineering team in the world. Arrogance would be too strong a word, but it was a form of unshakeable self-confidence that sometimes bordered on it. Nakamura-san had earned respect not through mere academic qualifications (although he was of course a graduate of a prestigious Japanese engineering university), but by decades of building up a superb technical decision-making batting average – proving to his peers that he was right far more often than he was wrong, that he could engineer and deliver vehicles, and hence that his judgement could be trusted. The final superpower was a rare one, in my experience, for an automotive engineering executive – he was an excellent public speaker. Engineers that can hold an

---

[13] *Both men and women in Japan regularly carry fans in the summer. Nakamura-san used his as a polite pointer – it being very rude to point with a finger in Japan, to emphasise a point by gently tapping on the desk, or occasionally to threaten to rap me on the knuckles; which he never did, of course.*

audience's attention are rare indeed, at least in the car industry. I was to see Nakamura-san do it many times – in Japan, in the UK and even in France. Nothing flash, nothing showy – he just had natural charisma. People wanted to listen to him when he spoke. My team adored him, and I had to quickly accept that *he* was always going to be the real boss – the best I could hope for was just some dimly reflected glory. But that was fine by me.

The first two weeks were hectic, while I settled down as Nakamura-CVE's left-hand man and started to get into the details of the project. We had the usual teething problems – due to some administrative snafu, no English-language computers or CAD 'tubes' had been ordered for my team. I kicked up a fuss, immediately making a reputation as an over-demanding upstart, and managed to beg, borrow and steal enough equipment to get things going. I even authorised my engineers to go to the local electronic stores (Japan is not short of these) and simply buy laptops on their own credit cards, to claim back later on expenses. This was not the done thing in Japan – breaking, as it did, many rules. In my very first days in NTC I had a rather uncomfortable meeting with the resources management team, who were trying their best to resolve the situation, but were not making things happen fast enough for my (admittedly very impatient) tastes. I made it very clear that I had a job to do, we had a vehicle to design, and that I would allow nothing and nobody to get in my way – certainly not some administrative process. I probably came across as an arrogant and impatient idiot, but a rather determined one. This reputation quickly spread ... and was to become both a help and a hindrance.

Of course, we had the usual issues when 25-odd people – some of them with young families in tow – move continents in a hurry: issues with visas, bank accounts, residence permits, company apartments, TV licenses, unexpected police encounters, new and sometimes over-possessive girlfriends. I will draw a discreet veil over these stories, but believe me that I had some very bizarre '*Lost in Translation*'[14] situations to deal with in our first weeks.

In between dealing with these 'events,' my core team and I started to absorb every possible detail about the B32A project itself. We did not quite have a blank sheet of paper – some work had already been done by our Japanese colleagues, and Nakamura-CVE duly assigned Kouichi-san, one of his project managers, to debrief me on progress so far. He did so with an eagerness that bordered on the slightly suspicious. He was quite naturally keen to off-load this project – Nakamura-san's team had many other projects, for Japan and the US, on their hands, and this Euro-car was far from being their top priority. They were almost visibly relieved to have this keen-as-mustard young *gaijin* ACVE arrive, to take over the least-important car

---

[14] *2003 movie starring Bill Murray and Scarlett Johansson. Murray's character bumbles around Tokyo trying to make sense of Japan - very much like my team and I in that spring of 2002.*

in the portfolio. We were very far from Europe here – and if Nissan Europe wanted to survive, well, it was up to us Europeans to take over this car and make it fly. Which was absolutely fair. So the core team and I spent many hours with our new Japanese counterparts as they downloaded reams of information – layout diagrams, provisional part BOMs (Bill of Materials), initial Concept Sheets with the first proposals for part-by-part concepts (mostly in Japanese, unfortunately) and of course, the crucial cost and profit projections.

The latter sounded the first alarm bells in my head. B32A was a very logical concept. The idea was sound, on paper. It was to take two existing cars – the Almera hatchback and its high-roofed MPV sister, the Almera Tino, and replace them with one 'medium-roofed' hatchback – a sort of hatchback on stilts, a little like the later Seat Altea. The fact was that both of these current-production vehicles were deeply unprofitable. They were not at all bad cars, but they were very much 'engineers' cars' and the result of extensive technical benchmarking of the competition – taking the best bits of a Ford Focus, a Renault Mégane, and a Toyota Corolla and mixing them up. The result is a perfectly good car, but one severely lacking character. Buyers unfortunately voted with their wallets, and stuck faithfully to their previous Focuses, Méganes and VW Golfs. As a result, Almera's COP was deep into negative two digits – I remember the shock I experienced when I first did the maths and figured out that we were effectively giving away approximately £1500 with each Almera that rolled out of the gates at Sunderland. It was literally as if we rolled up 150 crisp new ten-pound notes and tucked them neatly into the glove-box of each car, a nice present for each customer. Obviously unsustainable – we had to dig ourselves out of that profit hole, and fast. The theory of how we would do that with the new B32A was sound – again, on paper. We would cut investment costs by developing and tooling up not two body styles, but just one – a huge saving. We would try to design it so that it appealed to both market segments – the Golf or Focus hatchback buyer, but also the families with young kids that were still buying Mégane Scenics and other small MPVs in their droves. The vehicle was to be based on a Renault platform, allowing us to further save investment by taking 'off the shelf' parts developed by our Alliance partners, which would be already tooled up in Europe. The only fly in the ointment was NMISA – the car was planned to be built only in NMUK, while the Almera Tino was currently built by our colleagues in Barcelona. But even that was manageable – there were already plans in place for various 4x4 and light commercial vehicles to fill the production capacity in NMISA. So our colleagues in Catalonia were covered for the foreseeable future. So far, so logical.

Except that the numbers simply did not add up. Roger Blakey and I pored over endless Excel spreadsheets and what Nissan called 'Manhattan Skylines' – financial waterfall charts comparing the cost-and-profit start-point of

Almera to the equivalent projections for the new car. We reckoned that we could sell the new car at a sticker price a little above Almera. The logic for this was debatable (would a customer really pay more just for a higher roof?) but we persuaded ourselves to just go with it for a while. However, that still meant that the cost of actually building each car would have to be significantly less than that of building an Almera, and that was just not the case. The costs stubbornly kept adding up to a few percentage points below Almera, and while we could *just* get the COP numbers to be vaguely break-even, we were still far off the 6 to 8% positive COP that Ghosn-san demanded from all new projects. I started to worry ... but tried to shrug it off. Maybe all new projects started like this?

My misgivings crept back after my first visit to the Nissan Design Studio to see the initial sketches and models of the vehicle. The Design Studio was a secret-within-a-secret – the *sanctum sanctorum* of NTC. Kouichi-san escorted me there as part of our hand-off – along endless corridors, over a connecting bridge between two buildings, through various badge-activated doors, to a final automatic door where one final special badge gave one access to the bright lights, deep carpets and smiling receptionists of the world of Design. The air actually smelled different to the slightly-musty air of the dour Engineering offices – a pleasantly earthy odour hung in the air, coming from the clay used to build the various models. I was taken to the part of the studio dedicated to our project – yet more closed doors and screens to prevent even internal eyes from seeing other new projects. I walked into a brightly-lit space where there was a steel flat plate and sitting on it, a full-size clay model of 'my' very first project. A pretty special moment ... one of those moments you dream of when you are a kid dreaming of 'designing' cars.

My heart sank. It was, how can I say it politely? – boring. The clay model was asymmetrical, split down its lengthwise axis and styled a little differently on the left and right sides. A clever method that allowed one to assess two designs with one physical clay model, by the simple expedient of sliding a large mirror up to the centreline and making it appear like one, whole car. I walked slowly round both sides, as the designers and clay modellers stood around smiling, waiting politely for the new ACVE's feedback. I lied like a coward, smiled at them and muttered "*Sugoii! Iie desu.*" (Wow! That's nice.) or something equally insincere of the sort. Big smiles and bows. In fact, it was not 'nice'. It was certainly not ugly or badly executed. It was simply neutral. It was a competent styling exercise, making a hatchback a bit taller. Nothing stood out, nothing was shocking. This was my first time assessing a completely new clay model, but I have since had the privilege of seeing many such clays, and I have learned that it is important to have a strong initial reaction – either 'I love it' or 'I hate it' or 'What the hell is *that*?' In fact, some of the best designs are in this latter category – designs that perturb at first glance, that give one an unsettled not-quite-sure-about-this feeling in the pit of the stomach. It turns out that with time, and as your senses attune to the shape, these can often be the most successful. But it's important that you feel *something*. And with B32A, I felt almost nothing. It was like going on a blind date and feeling, well, totally indifferent. Something was missing. I felt the first inklings of panic as Kouichi-san and I made our way back to our office area. But once again I forced them down – this was my first project and I was sure as hell not going to give up on it before we'd even got started. I reminded myself that I was a beginner at this game – these doubts were no doubt normal. My blind date would grow on me. Right?

So, to work. For the next six months we worked like I'd never worked before. I was new, driven and under pressure. We so badly wanted to show what we were capable of. My team were the same – young, hungry, tireless. So started the rounds of what Nissan calls 'Concept hearings' – where each Engineering function presents its proposal for each component of the car – the basic principle, whether it is all-new, or to be carried over from an existing model, what shape it will be, what materials will be used, its mass, how it is to be fastened into the car, how the proposed design will guarantee its function and reliability, and – critically – what it is projected to cost, both in fixed costs (the up-front investment in engineering manpower, tooling, prototypes and testing) and in variable cost or per-part 'piece' cost. When I say each part, I really mean it – over 3000 individual components, divided into maybe 800 or so grouped 'systems', each system presented individually by the engineer responsible for it, usually accompanied by his senior engineer and manager. These Concept hearings – and every carmaker has its own equivalent – would take two weeks of solid work, the actual hearings or presentations taking all day, from 9am until 8pm, followed by hours of analysing the data from that day's hearing. My wife set up an automatic e-mail reminder from our home computer to my work PC – at 10.30pm every night, I would see a message from her ping into my inbox: 'Come home. Now.' I would do my best to do so, and arrive home, exhausted, somewhere near 11pm, eat something, collapse into bed. Next day, rinse and repeat.

Roger and I ground through the masses of accumulated data, desperately looking for ways to reduce the costs of this vehicle. And we started to find out just how hard it was. There was often an ostensibly good reason to carry over parts from a previous car, which meant that its cost was effectively fixed. Nissan has very strict design standards, and our Japanese colleagues were very careful that they be fully respected, even when we could clearly demonstrate specific cases where these standards made us uncompetitive with other European products. To my surprise, I found that some of the old, pre-Alliance Nissan ways were stubbornly hanging in there, three years after the deal was signed. Even though the *keiretsu* of allied suppliers had in theory been dismantled by now – one of the key actions of the Nissan Revival

Plan – I found that many parts had nevertheless been semi-automatically sourced with Nissan's traditional partners, rather than put out for genuinely competitive bids. The Renault-Nissan Purchasing Organization or RNPO was still in its infancy – although there was now, on paper at least, only one purchasing organisation, we found that old habits died hard – the Japanese ex-Nissan buyers still favoured their traditional partners, and had a certain disdain for the European suppliers naturally favoured by their – mainly French – ex-Renault colleagues. There was still a certain, unspoken 'this is how we've always done it' attitude in the teams. Engineers always had a very good reason why we could not shave a cent or two off any given part.

Worse, almost half the car was virtually 'fixed price.' The platform and powertrain systems were mostly carried over directly from Renault, and sold to Nissan at a fixed transfer price – a very common practice when car companies share underpinnings. And it was a take-it-or-leave-it price – I found that we could not really influence the cost of (say) the air-conditioning compressor or shock absorbers, as they were effectively to be bought from Renault, and for Renault-sourced items we could not easily change either the technical specification, or the price. Frustrating.

On the other hand, I worked closely with the Product Planning and Program Director's office to see if we could improve the cash-in – basically, how could we sell more cars, or sell them at higher prices? The PD was Imai-san, a formidable, very experienced Program Director who could be slightly intimidating, but seemed to have a soft spot for our little team of European refugees. He assigned his two best deputies, Iwamoto-san and Goto-san, to work with us and see how we could drag this project into profitability. We worked crazy hours, weekends and holidays, desperately trying to get the project to the minimum profit level we needed. Imai-san and I even went on an emergency trip to Paris to ask support from his counterpart in Renault, a certain Carlos Tavares. Note the name – we'll come back to Mr Tavares several times in this story. But Tavares-san had enough challenges of his own in managing the vast Mégane family of cars for which he was responsible, and was not able to help us solve the fundamental issues of our project. He did however very generously bend the rules by providing me access to a part-by-part cost reference table for the Mégane – this was to prove immensely valuable later on.

Winter came in Japan, the temperatures dropped and the green hills outside my office window turned sparkling white. By now, Nissan top management had got wind that all was not well with the European Almera replacement. Increasingly, Nakamura-san and I were called upon to report progress to various committees – Engineering, Product, Program. Brows were furrowed, and the obvious questions asked by concerned executives – can we find more volume? Is the vehicle pricing as stretched as it can be? Can we generate more revenue through more attractive options? But always, inevitably – "Dave-san, please try harder to find more cost reductions, *neh?*" I would bow, say *"Hai!"* (almost impossible to translate – the word does not mean 'Yes!', but something closer to 'I understand!' or 'Got it!'), promise to redouble our efforts, and go back to my exhausted guys to ask them to find me another 1% savings. Roger and I prowled the corridors of NTC like piranha looking for injured fish, pushing for any and all savings that could be scrimped here and there. I met with dozens of the proposed suppliers, cajoling, threatening and outright begging for cost reductions. We had some successes, but lots of failures. And the numbers stubbornly stalled – we were not even close to the minimum COP required by Ghosn-san. It had to come to a head. And it did.

Friday 13th December, 2002. I was sitting in the Nissan Boardroom, on the top floor of Nissan Global Headquarters, then based in the posh Ginza area of downtown Tokyo – it has since been relocated to Yokohama. The entire Executive Committee of Nissan were gathered – mainly locals, but with two well-known non-Japanese faces among them. One was Patrick Pélata, the product genius, head of Product and Programs. And at the head of the table, the already-famous face of Carlos Ghosn, the black upside-down-V eyebrows clenched in a frown of concentration, a dark shadow of beard showing that it was already late at night for this last meeting of the week. I sat on a second row of chairs, a respectful distance from the polished dark wood of the boardroom table itself, between Imai-PD and Nakamura-CVE. I was the most junior person in the room by many ranks. The Executive Committee had already been in session for hours – ours was the last topic on this Friday night – Agenda: go/no go for project B32A. It was decision time – 'go' would mean that my team and I would get to carry on our work and deliver this car that we'd already slaved over for nine months. 'No go' would mean a return to Europe, ignominious failure, probably the end of NETC and quite possibly of Nissan Europe. No pressure. Fortunately, I am not superstitious – the fact that it was Friday 13th did not increase the tension.

The meeting was called to order and Imai-san stood up. He calmly outlined with typically Japanese understatement that the profit numbers were not good, but that the vehicle was important to Nissan Europe and he would respectfully ask the Executive Committee for some flexibility in applying the minimum profitability requirements. He then said that he would ask Dave-san, Nakamura-CVE's deputy, to briefly take us through the vehicle cost status. Dave-san duly stood up and presented the facts on a few transparencies. Yes, even in late 2002 we still used 'acetates' – slick PowerPoint projections were still a few years off. I was nervous – this was the first time I had presented to the Executive Committee, the first time I had met Mr Ghosn in person, and the stakes could not have been higher. When I was finished, Ghosn-san thanked me, asked me a few detailed questions, then formally thanked Imai-san and Nakamura-san. He then said, "Well gentlemen, this is clearly an important decision for Nissan Europe, so I would like to have each of your opinions, please".

Very Ghosn-style. Each executive then got the chance to argue – briefly – for or against the decision. It was not unanimous. Some executives argued for leniency – that they should give the project a 'pass,' as otherwise Nissan Europe would die on the vine through lack of product. Valid arguments were made about safeguarding employment in NMUK and in the European dealer network. Others were more categoric – reminding their colleagues that this was exactly how Nissan got into trouble in the all-too-recent past, by not being disciplined in insisting on solid profitability. Patrick Pélata was one of the last to speak. I held my breath. He was Imai-san's boss, the head of Product and Programs, and everyone knew that he was Ghosn-san's *de-facto* right-hand man. His vote would weigh heavily. "Mr President, it pains me to say so, but I think it's a no-go. Imai-san, Nakamura-san and David's engineering team have done their best, but it's simply not profitable. Nissan Europe is dead anyway sooner or later, if we can't bring profitable products to market. We have to take the hard decision." My heart sank. And so it was. Ghosn-san, as Chairman, had the last word, of course. He efficiently summarised the general consensus and pronounced the Executive Committee's formal and irrevocable decision – no go. B32A was dead. He stood up, nodded kindly to me, and he and the other Executives filed out of the Boardroom to head home for some well-deserved rest over the weekend. The dream was over. We had failed.

CHAPTER 5

Digging in

I sat there, devastated. The very next evening – Saturday – was our team's Christmas party. We had booked a function room in a local hotel and our team was going to have a traditional Christmas dinner in a land very far from turkey-with-all-the-trimmings. All the team's families had been invited. It was an opportunity to let our hair down after a very stressful nine months, to thank the team and their families for their efforts, and to look forward to the New Year. I would, of course, have to make a speech of some kind. Except now, all I had to say was 'It's all over, folks, we fucked up. The project's dead and we're all going home.' Hell of a Christmas message to deliver. I felt faintly ridiculous as I felt my eyes pricking with tears – I was a grown man, this was business, and here I was, sitting in Nissan's Board Room feeling like I might actually weep – what an idiot!

"Dave?"

I looked up. It was Patrick Pélata, standing over me and holding out his hand. I stood up quickly and shook it. There were only a few people left in the room now. Patrick had spotted how hard I had taken the decision, had stayed behind, and walked around the huge boardroom table to try to cheer me up.

"I am really sorry. I know you've put your heart and soul into this. But we had no choice. You understand?"

I nodded that I did, too miserable to trust my voice.

"Tell me, do you still have your team of British guys here?"

I nodded the affirmative.

"What are they going to do next?"

I mustered up a wry smile. "Well, nothing, now. We were here just for this project. I have to tell them Monday that we're heading back to the UK, I suppose."

"Hmmm. I see." He bit his knuckle and thought for a moment, brows knitted behind his wire-rimmed glasses.

"What do you know about the 'Affordable Crossover' project?" he asked. I replied that I had never heard of it.

"Okay. It's something I've had my advanced Product Planning team looking at for a while now. It's pretty straightforward. You know Murano, right?"

Of course I did. The Nissan Murano was a mid-sized SUV that had been launched in the US several months earlier, in May of 2002. A good-looking, sporty vehicle, it stood out from the more traditional 4x4 'trucks' in the US, was well-priced and had great performance with its torquey V6 engine. It was a smash hit – we were unable to build them fast enough to keep up with demand. Everyone in Nissan knew this red-hot product – one of the first of many hits under Patrick's leadership of the Product team.

"Well, this 'Affordable Crossover' idea is very simple – it's an 80% scale photocopy of Murano for Europe. Same idea. Not a 4x4, not a hatchback – something different, in between. Can your team start to look at it from Monday?"

I swallowed – a man hearing the rattle of the falling guillotine blade, only for it to stop just as it brushes the hairs on the nape of his neck – and stammered out a too-quick "Yes. Yes, of course!" Patrick turned to my boss, still standing patiently next to me.

"Nakamura-san – is that okay, if David gets his guys to work on this project from next week?"

Nakamura-san nodded. "Of course. No problem."

"Okay, that's great. Dave, come report to me in a few weeks, would you? I'd like to see what you think of it, get your recommendations. Have a good weekend, okay?"

A final, typically French, handshake, and a kindly pat on the shoulder before he followed his fellow executives to the door and the weekend. Patrick knew damn well that he'd saved both my team and myself, but was far too much of a gentlemen to make a fuss of it. Neither he nor I could have known it, but he had just sown the seeds for what would become Nissan's fastest-selling vehicle, one of its biggest-selling, and the vehicle that would reverse Nissan's fortunes in Europe.

But things were still very serious. Our project was dead, and this new straw proffered by Patrick was just that – a single straw for a team that was still drowning. It was just two words for now: 'affordable' and 'crossover,' and it was merely an advanced study – one of the dozens that existed at any one time, the vast majority of which would never see a production line. I knew that I would have to send most of the team 'home' to the UK – and that they would be devastated. So, I still had a heavy weight on my shoulders as I headed back to home and the ill-fated Christmas party.

The party was great – for almost everyone. I had confided in my close lieutenants – Matthew, Roger and Peter – that B32A was dead, and we decided together not to tell the team until Monday. So we tried our best to be cheerful. We joined in the obligatory karaoke – Andy Palmer, who was by now himself in Japan, did a memorable rendition of the Sex Pistols' *Anarchy in the UK*. I made a special point of going around each table to thank my team's families, girlfriends, kids etc – knowing that on Monday I was about to disrupt their professional and private lives – again. I was miserable and felt both a failure and dishonest. The photos of that night look great – another fun party in Japan, with no visible sign of the disaster that was not merely impending, but had, in fact, already happened.

Monday morning, 8.30am. I called a full team meeting – a rare event. Everyone was there – maybe 30 people, including a few visitors on business trips from the UK, all standing up, elbow-to-elbow in a too-small meeting area near my desk. I broke the news – one of the hardest things I've ever had to do at work. The announcement of a cancelled project might seem trivial to the reader, and in fact it is – later on, I would experience other cancelled projects, and would grow a thicker skin. But back then, in these circumstances, it was different. It's hard to explain just how invested we were in this car. There were several factors that accounted for this. It was our first rodeo, as the Americans say. All of us were young and green. It was also high stakes – not just another car, but a car that had to dig Nissan Europe out of its profit trough and restore it to a viable business. We were foreigners far from home – isolated by distance, experiencing the curious 'long-distance greenhouse' effect that makes everything seem more intense. And I was probably guilty of semi-deliberately ratcheting up the pressure – emphasising if not exaggerating the 'do or die' nature of the project in order to extract the maximum performance from the team. But now all those factors came home to roost.

I was doing my best to stay cool and professional – conjuring Nakamura-san's quiet authority as best I could while explaining the decision and the rationale behind it, being very careful to fully support the decision and the Executive Committee position. No possible hint of playing the victim here – no 'guys, those top management bastards have stabbed us in the back' stuff. I lacked experience, but I was not stupid – I knew that this was just not the right way. But as I scanned the dozens of faces before me, I could read their emotions. These were engineers – British engineers for the most part. They were not people given to public displays of emotion or histrionics. The signs of distress were subtle, but obvious to me. Their eyes fell to their shoes. Feet shuffled uncomfortably. I saw their features collapse in disappointment; shoulders dropped as nine months of slog evaporated. A few muttered quietly to their buddies. It was bad.

The question session was tough – hard, pertinent questions. It would be dangerous to bullshit them – they knew me, and they knew the facts too well. So, I did not pull my punches. I underplayed the life-raft that Patrick had thrown us. I explained that the majority of us would be heading home, to NTCE, but that I would need some of the team to stay and help me start early studies on a potential new project. Lots of questions about this mysterious new project, and about who would stay. I deflected both, holding up my hands and asking them to be patient for a few days. They got the message, did not push me too hard – they could probably see how badly I was taking this myself – and dispersed back to their sections, heads down.

So this was it. Bottom of the hole. Nothing to do now but climb out. Flanked as usual by my solid wingmen, we started to rebuild.

I quickly presented myself to Advanced Product Planning, explaining that Pélata-san had mandated me to take on the 'Affordable Crossover' project. A few quizzical looks – who the hell is this guy? – a couple of calls to Pélata-san and doors opened up. I soon met Akihisa Suzuki. Suzuki-san was the Chief Product Specialist assigned by Patrick to work on the project and – now – to work with the remnants of my European squad and me to see if we could make it fly. Suzuki-san was that rare pearl – a product planner that knew what he did *not* want. It is all too easy for product planners – whether they be in the automotive industry, software or consumer products – to simply fall into the 'more is better' trap, and to assume that the customer just wants more of everything – performance, size, features, gadgets, options. This is what I call plus-plus product planning, and it's sadly, all too common in many industries. Great when you are working on an unlimited-budget supercar or luxury limousine, very dangerous when you are trying to offer a car to normal people, with tight budgets and other demands on their income – housing, education, healthcare. But Suzuki-san was one of those very few product people who could prioritise – he had a very clear vision of what the customer needed, wanted, and – crucially – a good grasp of what the customer did *not* need. This ability to concentrate on the 'just necessary' would be a key factor in the future project's success. He took me through the work so far, which was, thankfully, pretty much a blank sheet of paper this time. No real engineering or design work had been done – it really was a rough concept, just a few words on paper, describing the intended customer – an imaginary young European family with two kids – and the idea of a 'crossover,' as inspired by the Murano. I and my team had free rein to decide everything about the vehicle from a technical point of view – what platform, what powertrains it should use, the layout, supplier selection – everything. We had a second chance – and this time our hands were free. I felt a real glimmer of hope.

This positive news was more than offset by the unpleasant duty of going around the team, one by one, and telling them if they could stay in Japan, or if they had to go back to NTCE. This was hard – we had all planned to come to Japan for a year at least, and many people had disrupted their private lives to do so. Nobody wanted to quit the adventure now, when we were at our lowest point and clearly had unfinished business. But it had to be done – I sent about two-thirds of the team 'home' to work on more mundane tasks like cost reduction and maintenance of the current production vehicles, keeping just the folks we needed to do the core early work in outlining the new vehicle.

The next three months were intense. First, we had to decide the very fundamentals – how big would the car be? On what platform should it sit? What powertrains would go under the hood? This 'platform' decision is fundamental in modern automotive engineering. The platform is – loosely speaking, what we would call in the old days a 'rolling chassis.' It's the ensemble of the floor metal structure, the engine, gearbox, drivetrain, axles, brakes, steering and fuel systems – everything that is needed to make the car move and stop. Of course, a 'platform' never really exists in reality – the separate body-on-frame type chassis that really did make a 'rolling chassis' disappeared for most passenger vehicles in the 1950s.

In the 1980s and '90s, platform-sharing became a key weapon in the armoury of all competitive car companies. Instead of designing a new platform for each vehicle, platforms would be engineered to suit several different 'upper bodies' or 'Top Hats' – one platform could be used to build (for example) a five-door family hatchback, a four-door sedan or saloon car, a 5- or 7-seater MPV, a two-door coupé, possibly even a light 4x4 off-roader. In this way, the enormous costs of developing a platform/powertrain combination could be amortised over not just one vehicle, but five or more product lines, greatly reducing the costs to the carmaker and thereby maximising profits. This is exactly what Renault had so cannily done in the 1990s with the Mégane family – at one stage building seven different vehicles on what was essentially the same platform. It is what Nissan had signally *failed* to do prior to the Alliance, allowing our obsession with Toyota to lead us into creating more and more bespoke platforms, and allowing our costs to snowball. Correcting this, as we have seen, was one of the key actions of the Nissan Revival Plan.

Therefore, what was left of our little team started to consider what platform we could use, and what already existed in the Nissan platform 'catalogue.' First, we looked at using the Nissan B-platform. The 'B' segment is car industry jargon for superminis – the family of cars pioneered by the Ford Fiesta in the 1970s: small, compact city cars like the Fiesta itself, the VW Polo, Toyota Starlet/Yaris or Nissan's own Micra. The B-platform has several advantages – it was cost-effective, and was already tooled up in Europe, as the Micra was already built in NMUK. We duly looked at how far we could 'stretch' the platform – elongating the wheelbase, widening the track, and looking at the most powerful engines that we could squeeze under the hood. This new project was intended to be C-segment size – closer to a VW Golf than to a Polo, say 4.3m (14ft) in length – so using the B-platform meant that we would have to muscle-build it to the limits of its technical capabilities. After a few weeks we gave up, as we butted up against the platform's technical limits in terms of engine power and torque, wheel size and fuel tank size. It was also tricky – not impossible, but difficult and hence expensive – to convert it to offer a 4WD version, and we already had in mind that the new project would need a 4WD version to be accepted as a 'real' crossover and not just a hatchback in high heels.

So our attention turned to the 'natural' choice for a C-segment car – the C-platform. Here we had a tough decision to make. Nissan's existing C-platform in Europe was the ageing underpinnings of the Almera and Almera Tino. We quickly ruled it out – too old, too expensive, no 4WD capability, and it had never exactly set the world on fire on terms of ride and handling. Using the existing Renault platform – as planned for B32A – was, arguably, the logical choice. As we've already seen, it was already tooled up in Europe – it would not be too difficult to ship parts being used in Renault's plants in France and Spain across to the UK. But we ran into two obstacles – first, like the B-platform, it had no 4WD capability. Second, we had already seen on B32A that we would potentially have little or no influence over the part costs. The platform represents 45% to 55% of the total cost of a vehicle. We had to be able to control our own costs – completely. We could not therefore, take the whole platform 'as is.' But this time we *would* make sure that we used the available Alliance parts catalogue in Europe. Carlos Tavares had opened the door for us here in sharing the cost structure of the Mégane. I could see that certain components were far, far cheaper than we could ever source them using Nissan's feeble European volumes alone. We would be deeply pragmatic this time around – we would unashamedly raid the Renault parts bin and supplier base, and if that upset some of our more conservative Japanese colleagues, still stuck in a 'Not Invented Here' mindset, well, tough.

It quickly became apparent that we needed a new C-platform – more modern, 4WD ready, optimised for this new breed of 'crossovers,' and aggressively cost-engineered. This was a big decision – engineering a new platform is a major undertaking, impacting many vehicles and production plants, and hundreds of suppliers. At the time of writing (2021), the cost of developing such a platform from scratch would probably exceed a billion dollars. A platform has a lifetime roughly double that of a single vehicle – let's say 10 to 14 years. This was above my pay grade. I took it to Nakamura-san. As CVE, this was very much his decision – the platform 'belonged' to him. I presented our arguments. We did not intend to design everything anew. The team had scoured the parts catalogue and there were some useful elements that we could scavenge and recycle, thereby controlling the investment. In particular there was a Japan-market-only C-segment MPV called the Lafesta. Our team had identified that we could re-use quite a few elements of the platform of this vehicle – a lot of the floor metal forward of the B-Pillars, most of the front suspension, some engine bay parts. Even some elements of the rear floor unit could be carried over. Nakamura-san listened carefully, appreciated our frugal approach and gave us the green light to study further.

The plan slowly came together. A 'new' C-platform design emerged from the CAD tubes of the layout teams, my little skeleton crew working intensely, hand-in-hand with our Japanese colleagues. I remember many late nights spend poring over half-scale layout drawings draped over the chest-high drawing tables that still existed back then, marking them up by hand with Tim Dunn, our brilliant lead packaging engineer. Slowly a plan emerged to build not just one, but *four* vehicles on this new platform – and not only in NMUK, but in plants in the US, in Japan and in France. The economics made sense – and we started to have the bones of a project. We held what Nissan then called PM#1 or 'Project Meeting Number One' – a typically unromantic Nissan name for the first official management review for a new project. The plans were presented to the various functions – Product Planning, Finance, Resources Management, Engineering, Manufacturing – including the all-important platform decision. Green light – everybody liked the direction we were taking. And hence, the vehicle was baptised. Again, no romantic or fanciful code names – just a pragmatic alphanumeric code, simply the next one on the list: P32L. We did not yet know that it would one day be called Qashqai[15], and that it and its successors would go on to sell more than 3 million vehicles.

We were back in business. The next couple of months passed in fast-forward. My own stint in Japan would come to an end in April 2003 – my working visa had been for one year only. There was a lot to do in my remaining time in this fascinating country.

First and foremost, we had to define what this new car – P32L – was to actually *be*. The Qashqai has often been credited with inventing the 'compact crossover' market segment that would come to dominate the European and many other global markets from 2007 onwards. In my opinion it does not really deserve this credit, much as I would love to claim it! As we've seen, it was itself inspired by the Nissan Murano. And in fact, at this very early stage of the project, my team and I thought a lot about the Honda HR-V. Yes, that odd, boxy little vehicle that carried the somewhat dubious nickname 'The Joy Machine,' written proudly across the tailgate on some models. Based on a Japan-market vehicle called the Solo, the HR-V had been launched in Europe in 1999 and was at best a mild success. I would credit *it* as being the first 'crossover' in Europe. We studied it in obsessive detail and noted that one of the reasons for its failure had been that it was never accepted as a 'real' 4x4. This was for an almost ridiculously banal reason. The 2WD version of the HR-V used a simple twist-beam[16] rear axle

---

[15] *'Qashqai' was a name quickly selected for the very first concept car that would be shown at the Geneva Motor Show in March 2004 – a car that bore only a passing resemblance to the final production version. The name, inspired by a tribe of Turkic nomads living in Iran, famous for their beautiful woven carpets, was a hit, and was retained for the production version, despite the fact that many people found it hard to pronounce and impossible to spell (the key is – no 'u').*

[16] *A twist-beam or torsion-beam axle is a rear suspension design that uses the beam itself – often in a 'U' section joining the two rear wheels – as a torsion spring, making for a very simple, compact, low-cost design, which is however hampered by not being a truly 'independent' suspension design. It is often used on smaller, lighter vehicles where costs are more important that handling characteristics.*

– to save cost, of course. But its high stance meant that the axle itself was visible under its rear bumper ... a straight beam of metal running between the wheels. It made it easy for critics to mock it as a 'fake Jeep' or 'not a real off-roader.' We stored that away in our brains ... we would have to treat the rear axle and rear bumper design very carefully indeed, in order to avoid making the same mistake.

In parallel, Peter Brown led the team setting the detailed development targets – hundreds of them. Everything from performance figures – 0-100km/h (0-62mph) acceleration figures, ride and handling targets, detailed ergonomic dimensions, crash safety targets, corrosion targets, repairability targets, down to details like the glove-box volume and the exact capacity of the cup-holders. To do this, we forensically dissected the 'competition.' But we had the lesson of the Almera fresh in our minds – simply taking the best ingredients of similar cars and mixing them together in one large pot often resulted in a dish that might be technically well executed, but risked tasting as bland as hell.

We needed to invent a completely new dish. To do this, we looked at the best family hatchbacks – we judged them to be the Ford Focus Mark One and the latest Renault Mégane (the 'B84' Mark Two model). We then took the best of the small 4x4s on the market – undoubtedly the Toyota RAV4. We looked carefully at a few vehicles that had tried to combine the best attributes of family cars and 4x4s – the Subaru Forester and the (then-new) Mitsubishi Outlander, but very carefully – we did not want to be *too* influenced by these rather utilitarian vehicles. And slowly, Peter and the team came up with the recipe for what we now call the C-segment European crossover, by intelligently blending elements of all these cars. It had to handle, ride and brake well, and also be quiet inside – like the hatchbacks. It had to have good ground clearance and the high-perched driving position that people appreciated in the RAV4 – but without the clunky mechanics and cheap plastics of that vehicle's interior. It had to have 5 EuroNCAP stars to reassure the parents of young kids that would ride in the back, like the Mégane had just achieved.

Crucially, we found that the position of the window waist-line was critical: too high, and it would feel like a hatchback, like you were 'sunk into' the car; too low, and it felt insecure, too exposed, like sitting in a shop window. I spent hours with the layout and Design teams, defining the precise height and 'rake' of the door-glass line to provide what we now recognise as a 'crossover' feeling for the driver and front seat passenger. Finally, and this was key – it had to be cheap enough for the parents of young families to afford. They could afford a price tag that was a *little* more expensive than a basic hatchback, maybe, but it had to be a lot cheaper than the RAV4 or Forester. This meant setting very aggressive cost targets.

We also kicked off the Design work. The initial creative work would be carried out in the studios in Japan, but the intention was to launch a design competition – Nissan had just opened a new European studio in the UK, in London's Paddington area, and the new NDE (Nissan Design Europe) team would get their chance to 'win' the right to design this new project, in friendly but very serious internal competition with their Japanese colleagues.

And the project had a new Design Director – a fellow European, no less. Stéphane Schwarz was Swiss and spoke several languages perfectly. I liked him immediately – he was friendly, urbane, and effortlessly stylish, as befitted his profession. He was already famous within Nissan for his design of the final version of the Primera. This car, as I have mentioned, was a commercial disaster, being the wrong product at the wrong time, with a particularly poorly judged powertrain line-up. But the shape of the car, as penned by Stéphane, was something special. Not everybody liked it, with its radical, simple curves and surfaces, as shocking in their own way as the first Ford Sierra penned by Patrick Le Quément back in 1992, and its revolutionary interior layout and its centre-mounted instruments. But nobody ever accused it of being boring – Stéphane was clearly someone who had a strong individual vision, and was prepared to take a risk. We were to have some heated arguments in the years to come, but I always liked and respected him, and we remain good friends to this day. Stéphane, like (Akihisa) Suzuki-san, deserves a lot of credit for the success that was to come. No matter how well a car is planned, or how well it is executed by the engineers, if the design is not right, it is doomed from the start.

As the days in Japan counted down at an incredible and seemingly accelerating pace, one major task remained. I had to secure the project for NETC. There was never an open discussion about whether the project was to be Case 2 or Case 3, but I was nevertheless aware that there were certain people who had lost confidence in the European team. The B32A project had been a failure, by any reasonable measure. And it had been ours for nine months. Rightly or wrongly, that failure had rubbed off on us. It made our Japanese colleagues doubt us – for understandable reasons. We were all beginners – me included. Did we really have the ability to run a whole project? Would it not be safer simply to revert to the previous arrangement and make the project Case 2? The NETC guys could still contribute – as we had on Almera and Primera – but surely it was better to leave NTC in control, at least until the production tools had been released in a few years' time? Especially given that the platform and powertrain work would continue to be based in Japan for at least another year. And at least half of the design work would also be in the Japan studio – just how could a team based out of the UK possibly co-ordinate all that? All the more reasons to let an experienced Japanese team in NTC run things.

I was aware that these – very sensible – arguments were being whispered in Nakamura-san's ear. And I was determined to counteract them. No way were we to come this far, only to have the project snatched – albeit kindly, and with the best of intentions – out of our hands. No way.

Late one evening, I sat down at my desk and opened a blank Word file to compose my thoughts on the best way to secure P32L for NETC as a Case 3 project. I looked at the ceiling for inspiration, thought about the ill-fated Christmas party, about Andy Palmer and his confidence in me, and typed 'Anarchy in the UK' as the title of my paper. I knew that to secure this project, we would have to break the rules of the game, as the Sex Pistols had broken the rules of the music business. We would probably offend people as much as the Pistols did – and still do. And yes, it probably *would* be anarchy when we brought the project 'home' to the UK in the next few weeks …

My tactics were simple. Convince Nakamura-san that we *could* remotely manage the Design activity that would continue in Japan. Technology had advanced in recent years to allow this to be done, using remote cameras and a secure data transfer line. Commit to being in Japan myself to chair all the platform and powertrain component and cost reviews, and to present the Progress Meetings in person – knowing full well that this would mean an absolutely hellish travel schedule for the next few years. And most important – simply never ask the question about Case 2 or Case 3. I had not yet heard the classic Silicon Valley phrase 'fake it 'til you make it,' but I decided that the best way to do the project was – to just do the project. I honestly do not think I ever directly asked Nakamura-san or anyone else what 'Case' the project was. We just took it on, and kept running as fast as possible, hoping to get so far down the road that nobody could catch us up and stop us. And guess what? It worked.

It worked because I was prepared to take risks, and people knew it. This was exemplified by one telling remark by Nakamura-san. About this time, Nissan Japan's uniform changed – for the better – from a rather sickly hospital-green to a natty grey, with a narrow red stripe across the chest. We all received our new uniforms, neatly folded in plastic with a little card included to explain the philosophy behind the new colours. They were chosen, explained the card, to complement the new Nissan Brand Identity, which was 'Bold and Thoughtful.' The grey background represented the 'thoughtful,' cerebral side of the company, while the narrow crimson stripe was intended to evoke our 'boldness' in taking risks. I was absent when the uniforms were delivered to our desks, but someone later reported to me that Nakamura-san held his new jacket up to the light, appraised it quietly, then said "Hmmmm. We will have to order a special uniform for Dave. All red, with a grey stripe. Bold, and thoughtless." The 'bold and thoughtless' tag stuck, and I accepted the joke at my expense with a certain pride. Indeed, I was prepared to go through hell and high water, and to take whatever risks we had to in order to deliver this one. Just let someone try to get in our way …

CHAPTER 6

Anarchy in the UK

I looked out over a sea of faces. Maybe 500 people had gathered in the test building in NTCE Cranfield, the only space large enough to hold the entire staff. I was standing on a slightly raised podium, next to a full-size clay model of our shiny new project, P32L. It was April 2003, and the model was fresh off the boat from Japan – as was my team of temporary exiles and I, ready to continue the project and bring it to production. The crowd looked at me expectantly. I felt slightly uncomfortable – I was aware that the three dozen or so of us that had been hand-picked to go to NTC for the B32A project were regarded as having been extremely fortunate indeed. There was a danger that we would be seen as an 'elite' of some kind, that had been trained up in Japan to do the real thing – design a whole new vehicle from scratch – while our less fortunate colleagues stayed back to work on more mundane things like cost-reduction and the endless detailed engineering actions required to maintain the vehicles still pouring out of NMISA and NMUK every day. This was of course not true – the folks in Cranfield had started to work on quite interesting projects like the coupé-cabriolet version of the little Micra while we were away – but perception can quickly become reality. I had to kill the seed of this idea before it could germinate, and pull the whole company together behind this new project.

I let rip. Completely unscripted and unrehearsed, I delivered one of the most impassioned speeches I've ever dared to let myself give. I'm almost embarrassed to recall it now. I reminded the upturned faces how close we'd come to company death in the very recent past, and that Nissan Europe was still in danger because of its poor profitability compared to the 'domestic' market in Japan and, especially, compared to the US. I told them with absolutely no exaggeration that this was our last chance to design a car that was not only technically good, but would actually make money.

I then did something quite risky. I made the kind of unsubtle appeal to patriotism that only a non-British person can get away with. I reminded them that we were in the UK, that most of them were British, and that there was no need to remind them of their great engineering heritage. But – I can remember my slightly melodramatic pause-for-effect, walking around the clay model to stare at the whole audience – the British car industry was pretty much dead. This was close enough to the truth in 2003 to be uncomfortable for the audience to hear. The historic 'British' carmakers *were* all dead, or as near as dammit. Rover, Jaguar and Land Rover were in lamentable shape. The heyday of the UK volume car industry seemed to be long gone, the battle lost to the Germans and French in Europe, and to the Japanese on a global level (the rise of the South Korean carmakers was still in the future). Even Rolls-Royce and Bentley – historic names redolent of crackling Spitfire engines and sepia-toned victories at le Mans – had been purchased by the German premium carmakers that had clearly won the post-WW2 battle in the market. The only really 'British' carmakers that remained were niche brands like Aston-Martin and Morgan. True, the UK motorsport industry was healthy, but the only other positive stories were the Japanese 'implants' such as ourselves – Nissan in Sunderland, Honda near Swindon, and Toyota near Derby. But until now, even these highly successful plants had merely 'photo-copied' vehicles designed and engineered mainly in Japan. As an Irish immigrant to the UK, and a keen student of automotive history, I admit that I may have been over-sensitive to the historical significance of all this, but I was prepared to milk it nonetheless. More walking, more staring. I outlined all of this, pointing out that nobody in Britain was currently designing a true, mass-market, high-volume modern production car. Nobody ...

"... except us lot. We've been given the chance to actually design a car, from the ground-up that will be built by the lads in NMUK. I'm not talking about some little niche rich-man's wagon here. I'm not talking about messing 'round with something that was designed in Japan. I'm not talking about some bloody tin box that will cost more to build than we can sell it for. I'm talking about a car that will employ tens of thousands and make a shitload of money. And this is it." I pointed at the clay, sitting resplendent in its silver wrap, just in case they had missed the point. I decided that that was enough shameless rabble-rousing for now, took it down a gear or three, asked everybody politely for their full commitment, and let them wander off back to their desks, probably thinking that I had gone mad from too many Japanese noodles while I was in NTC.

Now, I certainly do not want to over-emphasise my demagoguery skills here, but the NTCE team undeniably did an amazing job over the next few years to deliver P32L. I don't know if my speech helped, or just made me look like an idiot, but I have never before, and rarely since, seen a team as united by the desire to succeed, as focused on one goal. Helped no doubt by all the factors I have mentioned before – and also a tinge of fear – the knowledge that Mr Ghosn and the rest of the top management would not hesitate a second to close us down if this car did not succeed. And 'succeed' was very clearly defined – it did not mean simply getting to Start of Production, or building a technically 'good' car like Almera or Primera. It meant succeeding as a *business* ... making a car that would be highly profitable.

As I mentioned earlier, a lot of technical target-setting had been carried out in Japan, at whole-vehicle, system and component level. Now we got down to setting aggressive cost targets – part-by-part. This was easy in principle, gruelling in practice. With my Program and Product Planning friends – still in Japan for the most part – we had estimated the price point of the vehicle, a very aggressive 25,000[17] euros for the base car. This was pitched to be affordable to the key customers – young families with a couple of kids. A little more than a standard family hatchback like a Ford Focus or a Toyota Corolla, but not much more. It requires only basic maths to lop the dealer margin off that 'sticker price' number, protect the all-important 8% COP that we were targeting, and hence figure out how much we could spend on actually building each vehicle[18]. It's then quite a lot harder to decide if you are going to spend that money on 'fixed' costs – meaning engineering manpower, tooling, prototypes, testing etc. – or on the 'variable' costs – mainly the costs of buying the parts from the hundreds of suppliers that would be involved. It's a balance – the more you spend up-front on the fixed costs, the less you can spend on the parts for each car, and vice-versa. Getting this fixed vs variable cost balance right is crucial.

Over the next months we ground through this process – never an easy one. Inevitably, the targets set part-by-part were almost impossibly aggressive. Through technical review after technical review, I informed each engineering group that they could have X euros for part Y. Cue impassioned explanations of why they needed X+20%: that the equivalent Almera part Y cost X+30%, that a 10% reduction was already a great achievement on their part, a technical 'tour-de-force' that I should be thanking them for. I would explain in no uncertain terms that they had X to spend, and not a euro-cent more. These discussions could get quite heated ... and I was prepared to go toe-to-toe with anyone who resisted. But I had a secret weapon, almost literally under the table. I was able to benchmark the detailed costs of the Renault Mégane. I mentioned previously that Carlos Tavares – then the Program Director of the highly successful Mégane family of vehicles – had allowed me to have a single, highly-confidential, for-my-eyes-only paper copy (it being far too sensitive to transfer electronically) of the costs of every individual part on the latest 'B84' version of the Mégane, which had been launched the year before, in 2002. This document was kept locked in my desk drawer at all times, unless it was actually in my hand. I personally checked to make sure that every component on our vehicle – and there were several hundred lines on the printed-out spreadsheet – was at least at the

---

[17] All costs will be quoted in euros for simplicity, the euro having been introduced in 2002. The GBP-euro exchange rate in 2003 was about 1.45 euro to the pound. The base-model Qashqai would eventually cost approximately £17,400 when it was launched in early 2007.
[18] Sometimes known as COGS or 'Cost of Goods Sold' in the industry jargon – this is what it actually costs the carmaker to build the car. A closely-guarded secret, for obvious reasons – the competition would love to know this detail.

same cost level, if not lower, than the equivalent part on the Mégane. I knew that this was both technically and commercially possible (as Renault had actually done it). I also knew that our selling price point was pitched slightly above the Mégane price point. If we could cost keep the costs (or cash-out) to the same levels of the Mégane, and the prices (cash-in) a little higher, then we were sure to have a solid business case. Horribly crude, very basic ... but it worked.

The folks who were fortunate enough to work on the P32L project would doubtless confirm that the level of 'aggression' on the cost engineering was as high as a team can reasonably stand. I was relentless – like a man possessed. Any opportunity to save not just a euro-cent, but a hundredth of a cent – was pursued, obsessively. My zeal made me the butt of my own team's wry jokes. At one typical Concept hearing, Steve Groves, the senior engineer for most of the vehicle interior, was presenting his proposals for the instrument or 'dash' panel. I really liked Steve, and would give him more rope than most, but even he started to run out of slack as he repeatedly pushed for us to use a new plastic material that had less vinyl-chlorides in it than the usual material used to injection-mould these components in the mid-2000s. Steve was, by the way, quite right – the use of these materials was indeed restricted a decade later to reduce their impact on the environment. But this was in 2003 – nobody in the industry had yet adopted these 'greener' but far-more expensive materials. Steve eloquently argued the environmental case. I pushed back, asking for competitor data to indicate that other carmakers would also go in this direction – that we would not unilaterally be asking *our* customers to pay a premium. Steve insisted on the deleterious impacts of vinyl-chlorides on marine life. I countered by requesting to see even *draft* regulation that might outlaw these materials in the lifetime of the car. Steve waxed even more lyrical on the effect on the oceans and aquatic flora and fauna. Finally, the rope ran out, I (predictably) cracked, and came out with "Steve! This is not the car to save the f@&king dolphins, OK? We're using the same bloody plastics as every other car-maker uses in every other car. Move on." This remark became (in)famous and was endlessly quoted by the team to embarrass me, not Steve. Apologies to both – I actually really liked dolphins, and Steve, and I still do. When their costs fell, I duly adopted vinyl-chloride-free plastics on all future models I was to be responsible for, well in advance of the new European laws that Steve had rightly predicted – if a little bit too early. Every time I approved vinyl-free plastics from then on, I thought about poor Steve, and promised myself to try to be more patient.

Costs are one thing, but revenue is another. Costs are under the car maker's direct control, but revenue is anything but. Revenue is simply the number of cars you can sell – the 'volume' in carmakers' jargon – multiplied by the actual price paid – which, as everyone who's ever haggled for a new car knows, is not necessarily the 'sticker' price, but the real price paid by the customer, after any discounts, free floor mats, trade-ins etc. Predicting revenues is an art, or at best a craft, but it is most certainly not a science. Carmakers have great difficulty in forecasting the sales volumes of even the nth generation of a new car like the VW Golf that has been around for decades. Forecasting the sales volumes for a new concept like this P32L – neither a hatchback nor a 4x4 but something new, in between, and nothing at all like its previous siblings, the Almera and Almera Tino, was pretty much as scientific as crystal-ball gazing.

We'd set a fairly logical price point, as explained. But now the fun started – how many cars could we sell? We studied the sales of the Toyota RAV4, looked again at the Honda HR-V, and Suzuki-san of Product Planning took a brave decision. I remember him saying "Okay, gentlemen. I think we can sell a peak volume of 130,000 cars per year." I recall looking at him and thinking that he was crazy. Almera at its *best* year had sold 116,000 vehicles in Europe. The HR-V? Just 26,000. 130,000 seemed very ambitious for such a new concept as this 'crossover.' But Suzuki-san was confident. "Dave-san, you and your guys are going to engineer a great car, and you're going to hit the cost targets, so we can keep the price as planned. Stéphane-san is doing a great job on the Design. I've seen the lastest clays and sketches. They are great. I really think we can do 130 in the third year of sales. *Ganbarimasho!*" ('Let's try!'). I leaned back, and thought – why not? Spoiler alert here – Suzuki-san was dead wrong. Qashqai would go on to sell almost twice his estimated peak volume, and to hold that level for several years. NMUK would run flat-out, building almost a quarter of a million cars per year, year after year. He was – we all were – out by almost a factor of two. But in the right direction …

He was right about Stéphane's work, though – he and his Design team were doing an absolutely fantastic job. As I've explained, half of this work was being done in Japan, in the 'central' design studio back at NTC. But half the work was now done in Nissan Design Europe, in London, the two studios battling it out in a competition as we narrowed the design options down. Nissan used a similar approach to many other carmakers – a process of elimination, through stages called 'Go With Two,' where two clay models are selected from four or more proposals, then a 'Go With One,' where a final – single – model is chosen. That 'Go With One' model is then refined down to the final 'Model Freeze' design. Stéphane had now also moved from Japan to the UK, as the centre-of-gravity of the project – engineering, design and manufacturing – moved to Europe. I worked very closely with him, going very regularly to the London studio, and making sure to visit the Japan studio every time I was in NTC. We fought regularly – but in a good way. Chief Designers and Chief Vehicle Engineers rarely see completely eye to eye. There is usually – in fact there should be – 'creative tension' between the two. But if these tensions spin into conflict and animosity, the car is doomed. Two ingredients are necessary to keep the relationship on the right side of the 'creative' line. First, respect. The designer must respect the engineer, and vice-versa. If the Chief Designer thinks he knows more about engineering than the CVE, trouble lies ahead. If the Chief Engineer thinks he has better taste than the Chief Designer, expect a disaster. Second, they must have a shared vision of the final car. If the Designer is trying to design a Hummer and the Chief Engineer thinks it should be a supercar, well, the result is not going to be pretty, or good. Right from the start I recognised that we had both of these key ingredients in my relationship with Stéphane. I respected him enormously, and I never felt that he was trying to 'engineer' the car behind my back. We both had a common vision of the final car – and I trusted his taste implicitly. So when the sparks flew – and they did, regularly – it was never bitter, and we both knew we were fighting for the same cause: the good of the project. The design of the final product was generally acclaimed at excellent – not trivial, or flashy, but quietly understated and firmly establishing the visual language for a modern crossover – and that is Stéphane's achievement. If we ended up selling twice as many cars as Suzuki thought we 'should,' in very large part that is due to the work of Stéphane and his designers in Japan and in the UK.

Meanwhile, I was spending far too much of my life on Boeing 747s. The powertrain and platform engineering was still based at this stage in NTC – for the very good reason that these key elements would be shared by at least three other vehicles, under the control of the CVE. The principles of my own 'Anarchy in the UK' paper had committed me to being in Japan often enough to project-manage these engineering activities, as well as the 'Upper Body' and design work being led from Europe. I had promised Nakamura-san that I would be able to handle the challenge of the project being spread half-way across the world, and I had to uphold my promise. Nakamura-san was by now Nakamura-SVP or Senior Vice President, promoted into the upper ranks of Nissan management, handing his job as C-platform CVE over to Matsumura-san, who was now my boss. A huge motorsport fan and a brilliant powertrain engineer, Matsumura-san was a tough boss and came close to scaring me at times. I travelled to Japan – a 12-hour flight each way – over 14 times in one year between 2003 and 2004, staying each time for 10 days to two weeks. While I was in Japan I would meet with the powertrain and platform engineers, and exert the same relentless cost pressure on them as I did on my 'own' guys back in NTCE. I was as bull-headed and unpleasant to my Japanese colleagues in squeezing cost out of the car as I was with my European counterparts. I was also beginning to feel exhausted ...

There were also fun times, though. I did occasionally get to do what folks imagine the job of a Chief Vehicle Engineer to be – testing cars. We had to drive the competitor cars extensively. We carried out this testing at Nürburg in Germany – no, not on the famous Nürburgring, but on the public roads around the mythical Nordschleife circuit. This was an excellent base from which to benchmark vehicles, for several reasons. First, Nissan had a large workshop there, where we could discreetly store, maintain and analyse cars. Second, the roads around the circuit are excellent – a great mix of all types of road surface and weather likely to be encountered in Europe. The locals are extremely car-friendly – they are unlikely to complain about convoys of cars driving through their quiet, pretty villages. It is not far to drive to find stretches on speed-unlimited German autobahns, including one long stretch in particular that has a steep, even gradient, allowing practical testing of cars' maximum speeds on sustained slopes – something impossible in the UK, and very difficult even on closed high-speed bowls, which are generally flat. I spent many happy hours with Peter Brown and his team driving the benchmark cars that we had selected in Japan around the Eifel mountains – the Renault Mégane, Ford Focus, RAV4 and Subaru Forester, with occasional other vehicles thrown in for variety. It was a pleasure to be just a car enthusiast for a few hours, torturing these vehicles – albeit in a very structured way – around the Nissan testing courses, able to forget my endless lists of component costs and technical reviews for a few hours.

And slowly but surely, the vehicle morphed from paper proposals to digital data on the CAD screens, to metal and plastic and rubber. The first vehicles to be built were Frankenstein's monsters – what Nissan called 'PF mules' or platform mules – cut-and-shut cars based on existing models, cut apart and welded back together to provide the wheelbase, track and approximate suspension geometry of the new vehicle. The new powertrain would be installed under the hood on hand-fabricated engine mountings and bracketry, with some creative methods used to get it to sit in the right position. The body and interior are afterthoughts – anything will do, as long as it has roughly the right dimensions, mass, and allows the air to flow into the cooling and braking systems in roughly the right way. The first of these mules was built in Japan – at NTC – and was ready for testing in early 2004.

I travelled to Japan – once again – and made my way to our Tochigi Proving Ground, 100km to the north of Tokyo. Tochigi is exactly what you might imagine by the words 'proving ground'– a huge, top-secret car playground, with every conceivable type of test track, road, surface, obstacle, etc. Access was very tightly controlled. Even Nissan permanent employees were vetted, photographed, metal-detected, and subjected to endless form-filling to get past the turnstiles and typically strict but polite security guards. I duly jumped through these various hoops, and was accompanied by some of my Japanese colleagues to the little workshop where the very first 'P32L' mule sat. It was deeply unspectacular – a grey-and black creation that was loosely based on a Lafesta bodyshell, chopped and stretched in all dimensions. But under the skin it was a real P32L, with the little Renault 1.5L diesel engine, codenamed K9K, that would be offered – an unusually small engine of this type for the time. We chose to build this mule with this, the smallest, newest and least powerful engine that we would offer, as we were worried about its performance – would it feel 'peppy' enough? Would it be able to haul around a relatively big and heavy car in a satisfactory manner? Well, that was what I was here to find out.

Except, I couldn't drive it. Only qualified Tochigi test drivers could drive at Tochigi. I had the basic, entry-level Nissan test driver's licence, but not the more advanced level that would qualify me to drive a unique, one-off mule, on the twisting roads of the proving ground handling circuit. *Mondonai* (no problem); a test driver had been assigned to me, and he would drive while I sat in the passenger seat. I was of course disappointed, but did not argue – safety was safety, and in Japan, one did not challenge the rules – not directly, in any case. I bowed graciously, thanked my test driver, who spoke no English but smiled warmly at me, and strapped into the passenger seat with my little notebook on my knee to record my sensations in this, a pretty historic test drive. Off we went. My driver drove smoothly and expertly as all these guys do, inputs honed from tens of thousands of hours of wheel-time. Off across the proving grounds, through the labyrinthine approach roads and out onto the handling circuit.

He almost immediately pulled into one of the little concrete shelters that were dotted around the track to allow cars to hide from any low-flying aircraft or helicopters chasing 'scoop' shots of prototypes with long telephoto lenses. To my surprise, he got out and walked around to my side. He then motioned for me to get in the driver's seat *"Anata-no kuruma desu neh?"* with a huge smile – "It's your car, isn't it?." He knew that nobody could see us and that we looked pretty much the same anyway from a distance, wearing identical Nissan outfits and hard hats. This was Nissan rule-bending – officially I never drove the car, unofficially it was to be my honour. I did not have to be asked twice – I hopped into the other seat, bowed deeply to my now-passenger, and set off – very carefully – around the circuit. I admit that I drove like the most cautious of the reader's elderly relatives – now was not the time to crash the car, which would result in lots of paperwork and tricky explanations that would get my new friend in lots of trouble. But I savoured every kilometre and every gearchange, and noted that the little K9K Diesel was indeed torquey enough to propel us around at a reasonable rate. All went well, I kept the car on the asphalt, and nobody ever knew about our little rule-interpretation.

Through 2004 and 2005 the project progressed apace. I lived, breathed and dreamed P32L. Although I had by now been promoted to Director of Programs in NTCE and was hence now responsible for other projects, they were almost distractions. I was obsessed by the car, and genuinely often dreamed about it. I could visualise it in 3D in my head, every nut and bolt, every component, every curve, could walk through it and rotate in my head around three axes, like it is now possible to do with Virtual Reality headsets:

but I could do it then just by closing my eyes. I knew the cost of every part, not just to the euro, but often to three decimal points of euros. I felt more than I owned it – I felt it was a part of me. I knew the designer for every part, personally. For me they were not parts, they were Dave Kelly's rear roof traverse, Steve's cup-holder, Lance's trunk trim, Barry's seats, James' airbags, Grahame's air filter. I knew them, and their creations, intimately.

We made mistakes – lots of them. The next time you pull up behind a Qashqai in traffic, have a good look at the rear bumper. You will see there is a rear foglight in the centre of that bumper. It's an afterthought – a mistake. Originally the car was designed with just the two rear light clusters that you see on the body sides to either side of the tailgate or 'hatch' – with fog lights to be built into these clusters.

I was at my desk one day when two of the exterior parts engineers came to see me. Problem. They explained that they had checked the height of the fog lights, and they were too high. You see, European 'Type Approval' law requires the rear fog lights on a car to be no more than 900mm above the ground, to avoid any possible confusion with brake lights, which are mounted higher up. The engineers explained that they had very carefully checked, and they were 901.2mm off the ground – 1.2mm too high. We would have to re-tool the rear lights completely – hundreds of thousands of euros worth of tooling, even before we considered any knock-on impact to the body metal surrounding them, which would cost many multiples of this sum.

This was very serious – we had already launched the production tools for both lamps and the sheet metal. I admit that I lost my temper – I was young and fiery then. I probably hit the table, definitely used strong language and will have ranted and railed for a bit before calming down and trying to understand the error. It was simple, as these things often are. Cars have different 'ground planes' to represent the different lading conditions – fully laden is five people aboard, with their luggage – 500kg pressing the car hard down on its springs. 'Kerb' condition is the opposite – the car empty of all human beings, just containing its fluids and some fuel, sitting up on it springs, maybe 20 or 30mm (0.8 to 1.25in) higher than the fully laden condition. Layout engineers also simulate various conditions in between – one person in front, two in the rear, two in front, various combinations of luggage and fuel loadings. In short, this results in a whole plethora of ground planes. One of the engineers had made an honest mistake. Probably tired and bleary after another 12-hour day at a CAD tube, he had mislabelled a ground plane. Another engineer had unknowingly opened the mislabelled file and worked with it. Somehow, it has escaped all our careful checks and technical reviews. Result – the lamp was in the wrong place. The car was now illegal – it could not be sold in Europe. I went home and tried to let my body sleep while my brain did not.

Next day, we hashed out a fix in a couple of hours. It was too expensive and too long to retool the light clusters, let alone the massive tools that stamped out the bodysides. We would simply have to take the rear foglight function out of the clusters and add a new foglight in the centre of the rear bumper. True, this would mean throwing the bumper tool away, but that was 'only' 250,000 euros – an on-cost that we could absorb with savings elsewhere. That left the tricky job of explaining to Stéphane and the Design team that we'd need to make a later-than-last-minute change, and redo most of the rear bumper whose shape they had signed off months before. They took it in good heart, albeit with lots of ribbing – again, an example that the vehicle really was made by one team, working together. Helped by Stéphane, the suppliers, our Purchase colleagues and the whole wider team, we made it happen.

I could list many such examples – walk you around a 2007 Nissan Qashqai and point out all the little errors that nobody knows about but those who committed them. But the millions of happy customers forgave us, fortunately, so I will draw a veil over them. There were bound to be such errors – we were, after all, rank beginners. But slowly, through various design reviews, concept hearings, and prototype iterations the car came into reality. Suppliers were selected – not easy, with the ferociously-low commercial targets that we had set. Production tools were released one-by-one – many of which I personally signed off – and the various suppliers all around the world sprang into action – milling, welding, injecting, forging, machining and moulding the thousands of individual parts into physical reality.

Autumn, 2005. I'm back in my second working home – Japan. I had taken the London-Tokyo flight so often by now that I could pack my little overhead-storage bag in my sleep. I did not even bother booking a room in our usual hotel, the 'Royal Park' in Hon-Atsugi – they knew me by name and would simply take a set of room keys off a hook when I walked into reception. When I arrived at my desk in NTC, there would be a stack of documents and drawings waiting for me review and sign – as there was on my other desk, almost 10,000km away back in Cranfield. I was hot-desking between two offices on two continents, managing a team of hundreds of engineers spread between the UK and Japan, and it was starting to take its toll. But on this day I was not in NTC at Hon-Atsugi, I was a few stops down the train-line in Nissan's Zama plant. This was one of Nissan's historic plants, now converted to become a specialised protype build plant – building the first, prototype-tooled examples of each new model, before they were handed over to the production plants that would build them in mass production. And I was here to see the latest car they had built – the very first 'real' P32L.

I had unfortunately missed most of the build process itself, as I has to attend a management training course for a few days. But my team had been at Zama for about a week, working with the prototype build technicians, skilled specialists in hand-assembling the very first cars of any new programme. In final production the cars would be popping off the line every 55 seconds, but this 'Car Number One' might take a week or more to build, each part being hand-fitted, and often needing a little old-school fettling to fit – a hole enlarged with a hand-drill or a file here, a part 'encouraged' into place with a strategic tap of a soft-faced hammer there. Parts were often missing, late, or hand-delivered direct to the build by apologetic suppliers. The team had been working crazy hours for a week, keeping me informed by e-mail and telephone of the progress, cooped up as I was in a training centre far from Zama. But finally the damned training course was done, and I grabbed the first taxi and train that I could find to rush to the Zama plan – just in time.

The car was indeed finished, but had not yet been shipped off to Tochigi for shake-down testing. By the time I got there it was late in the evening, and the team were heading home for some well-deserved rest and a short night's sleep before tackling Car Number Two. I was relieved that my own guys were already gone – probably hitting the beers in some local bar. But there was one older build foreman there to look after me. I thanked him for staying behind – much bowing and *dozo*'ing (*dozo* – 'you are welcome' or 'please' in informal speech). He guided me through the almost-deserted cavernous old plant, to the build area where the completed car awaited, sitting surrounded by floor-to-ceiling plastic confidentiality shields, like a giant version of the curtains that one pulls around a hospital bed. The foreman pulled back a corner of the curtain that served as a door and invited me to help myself, while he bustled around tidying up the zone, shutting off power supplies and lights for the night.

I stepped into the area and saw the very first Qashqai ever built. It was white (most prototypes are – white helps in spotting defects like body cracks or corrosion) with unpainted black bumpers and ugly temporary steel wheels. It was a bit shoddy – with a few missing parts, gap- and shut-lines that were far from production-ready, sitting awkwardly too high on overly-stiff springs. But it still stopped me in my tracks. It is hard to describe how emotional it was to see this object for the first time. It had filled my mind for the last three years, obsessing me, driving my team and I close to our limits. And now, here it was – sitting in front of me, just as I had imagined it, just as I had seen it hundreds of times on paper drawings, clay and digital renderings, and in endlessly-replayed 3D movie projections in my mind. I admit now that I was quite moved. Luckily nobody was there to see or to laugh at my ridiculous over-reaction to a mere car, but it really was that overwhelming. The car had become a reality. A decade or more of effort was sitting in front of me, black-and-white and shoddy, but beautiful to me.

CHAPTER 7

Crossroads

As 2005 rolled on, and the relentless P32L master development schedule ground on with it, I started to feel the exhaustion set in. The excitement of the B32A project, followed by the crash to a halt had taken it out of me ... I had been too invested in that project, and had taken the cancellation too hard. I felt immense personal pressure to recover the situation and make P32L a success. Now, for the first time in months, I had felt the pressure lift marginally, following the successful first 'real' vehicle build in the Zama plant. This car *was* going to make it.

The various tasks progressed through seemingly endless 70-hour working weeks. We gradually worked our way through the critical supplier selection for all the parts – from the big, expensive, visible parts like the headlights, instrument panel and steering wheels, that cost hundreds or euros each and where the tooling ran into many millions, down to the tiniest and least glamorous brackets holding brake and fuel lines: little bits of bent steel that the customer would never see, that might cost 30 euro-cent and be tooled for less than the cost of a single car. The sourcing competition was fierce – and fair. Any supplier who 'won' the business on P32L deserved it – the cost targets were mercilessly low, and we gave no quarter at all in the selection – each one was done on hard, cold engineering and cost data. 'Traditional' Nissan-friendly suppliers that had supplied a certain component to NMUK for the last 20 years lost business to new companies if they were not competitive enough. Many suppliers that Nissan had not previously used but that Renault had used for decades were introduced. Old habits based on 'well, we always worked with so-and-so for these parts' were forcibly broken if necessary. It was a rigorous and no-holds-barred process. Once again, I signed off each and every sourcing decision, and participated in many of the 'last call' meetings to negotiate the last euro-cents off the final price of key systems.

The next physical step in the project, after the 'mule' or 'PF-Lot' car that I had (officially not) driven in Tochigi and the small batch prototype, hand-built cars that we had built in Zama, was to run the first Production Trial in NMUK. This was a major milestone, and we were now working intensively with the NMUK manufacturing engineers to prepare for this. Nissan – in common with most Japanese carmakers – was very strong on what is known as 'simultaneous engineering' – meaning that the production engineers who will be responsible for building the car for the next 5, 6 or 7 years, will work literally alongside the product engineering team to make sure that they not only merely *understand* the design, but give active input into it: making sure that all the little errors and design inefficiencies of the previous models (in our case, Primera, Micra and Almera) were corrected in the new model, making it as quick, simple and efficient as possible to build – hence improving its cost and quality for the end customer.

We were now deep in this process, and the offices in Cranfield filled with North-Eastern accents as the guys from NMUK came 'down South' to work with us. Chief Vehicle Engineers often have pretty fraught relations with the production plant, but I can honestly say that I did not – I always found my NMUK colleagues incredibly dedicated and very demanding, but always in a positive way. They would not hesitate to point out if the design engineers in my team had made an error – in no uncertain terms. It was not a working culture for the thin-skinned. But it was never vindictive, and it was always with the sense that we were one team, working for one goal. This was where that mandatory NMUK 'line training' all those years ago paid off – everyone on the team knew just how hard the job of the men and women working on the line really was, by personal experience. If we could make their lives easier by simplifying how a part was clipped, screwed or glued into place, we would. We remembered all too well the aching shoulders, bruised knuckles and sore fingers of our brief stint 'on the line' – and were aware that we were paid far more than our colleagues at the sharp end of the business for our relatively easy office jobs.

Not *that* easy, though. The travel and the long working hours started to take their toll on me. My wife was hugely patient with me, but I could see that even her patience was wearing thin – I was usually already gone to work before she woke up, and she rarely saw me before 10 or 11pm on a weekday. I was working even longer hours than I had in Japan. I tried very hard – and mostly succeeded – to ring-fence my weekends, but those efforts were largely negated by the travel to Japan. I went there at least once a month, for two years straight. My wife used to joke – half-seriously – that I actually went to Japan for two years, not one, but that I had just simply 'forgotten' to tell her. I would usually fly on a Saturday, stay two weeks, and come back on the following Sunday, to avoid travelling during the working week – effectively killing three weekends in a row, or half the 'free' time in the month. At the time it did not feel tiring – I was loving it, living on the slow-burn adrenaline that such intense projects generate – but with hindsight, it was undoubtedly starting to tell on me.

I was pushing the teams almost as hard. I was demanding to the point of being unreasonable. I would ask people to do the near-impossible, and I was unforgiving if the standards we had set ourselves were not met. But I held myself to the same standards, and tried very hard not to ask anyone to do anything that I would not – or could not – do myself. I was leading from the front, but in the manner of a crazy young second lieutenant that leads his platoon across a minefield – it works, you will probably get there, but you may leave people lying on the field behind you.

I remember the Body team manager coming to me at the peak of the activity of releasing the 'Body in White[19]' data – this is the computer data that defines the shape of each of the pressed steel panels that makes the body itself. Each 2D drawing and the vital 3D CAD data defines a panel in precise detail – not just the general size and shape, but the complex radii of each bend, the tiny rolled-over flanges that are invisible when the car is assembled, the position of each hole into which to clip an electrical harness or trim part, the precise location of each spot weld and each sealing mastic bead – enormous quantities of very detailed data, that needed to be micron-precise, as it would be used to mill the enormous press dies that would stamp each part, and the mind-blowingly expensive robots that would assemble them in the Body Shop at NMUK.

The Body Manager, Ueda-san, came to see me with the worrying greeting of "Dave-san, problem." He went on to explain that we were running about six days behind schedule for the non-negotiable Production Release date for the body data – the date when these data needed to be released to the die suppliers and the production plant. It was now only a month away. I did some quick mental arithmetic and looked up at him. "No problem, Ueda-san. There are three weekends left before the end of the month. Three weekends – six days. *Mondai nai* (no problem)." We walked together back over to the Body design area, called his team together and I explained that we would be working weekends for the next month. I did not ask, I simply announced it as a *fait accompli*. Faces looked up – exhausted faces, already drawn from months of this work. I was now telling them that they would work nearly 30 days straight – no weekends, no playing with their kids, no trips to visit family. We were late, and now we would fix it. Nobody complained. They just nodded, grimly, and got on with their work. Of course, I could not ask them to do what I would not do – so I came in too, every day, and bought the necessary pizza to keep blood sugar and morale up. We got the job done, and we moved on. The P32L body data was released on time and to the required quality – to the day. What a team.

As the summer of 2005 turned into autumn, and the Production Trials loomed large, I travelled again to Japan. Very unusually, I was asked to report

---

[19] *Body In White or 'BIW' is car-industry jargon for the stamped metal bodyshell, complete with its doors, bonnet (or hood) and tailgate or boot/trunk-lid, but before final painting or the addition of any of the 'bolt-on' components. There is some debate over the origins of the term – but it most likely comes from the light grey (white-ish at a push..) colour of the primer that the body is dipped in to protect it from corrosion.*

to Nissan Global HR in Headquarters in Ginza. I grudgingly made my way downtown by train, cursing the hours lost to the project, and duly reported to the then-head of Nissan HR. He explained that I was to receive a great honour. There was an opportunity for me to become CVE – to be promoted from 'Assistant' CVE, to the real thing – Chief Vehicle Engineer, with no qualifying 'A.' I would be the first non-Japanese person to have this opportunity. It was indeed a great honour. I knew well that the HR department alone was not responsible for such a proposal – it was a decision that would have been deeply discussed within the Engineering community, and that it would undoubtedly have been supported, if not directly suggested by Nakamura-san. I thanked the HR boss, and took the Tokyo metro and suburban train west, back to Hon-Atsugi and NTC, deep in thought.

Back at home in the UK, I discussed it with Cooleen, my wife. She knew how much this meant to me, what an opportunity it was, and told me to go for it. But I had misgivings, and the more I thought about it, the more my conscience pricked me. From a purely professional point of view, there was no hesitation. I loved Nissan as one loves one's first love. This was a huge opportunity – to be responsible for a whole family of cars on one platform, not just a single car in the family. I was still only 35 – very young, in Japanese company terms. I knew that this was a stepping-stone, as it had been for Nakamura-san, to the big-time. He had recently been promoted to SVP or Senior Vice President. After that was Executive Vice President – the very top of the tree in the Engineering world, reporting to Carlos Ghosn directly. I would be only a few steps down the ladder. Back then, I was ambitious and had limitless self-confidence. I genuinely thought I could make it to the very top of the Nissan corporate ladder. It was a hell of a temptation. But … I knew myself too well. The offer from Japan was clear. This was not a return-trip one-year or three-year assignment. This was a one-way ticket to Tokyo. I would work the rest of my career in Nissan Japan. I knew that it would absorb me, totally. I had seen the Nissan senior executives from up close, had observed their professional lives, totally dedicated to the company. These guys did not have hobbies, or interests outside the company. Possibly the odd round of golf – inevitably played with colleagues or ex-Nissan folk, or fellow executives from the suppliers. Their dedication to the company was all-consuming – closer to a religious vocation than a mere 'career.' I was not sure if I really had this level of commitment. And even if I did, I knew what that would mean for my private life. I had already seen many marriages and long-term relationships collapse through the pressure of the job – folks spending too much time in the office, too much time away from home, absorbed and obsessed by the company and the cars, to the detriment of their partners and kids. I did not want to become just another one of those statistics. It was also a hell of a thing to ask my wife to do – she loved Japan, loved the six months we spent there, but we had to face it – she was a blonde

Irishwoman who would always be an outsider in Japan. Not that I exactly fitted in – I was a blond and blue-eyed *gaijin* too, but I would have a defined place in Japan and in Japanese society: I would be a senior Nissan executive, with a position and recognised status that would give me a solid foothold in the county. She would always be simply 'my wife' – it would be very hard for her to restart her career and carve out her own place in this beautiful, welcoming but oh-so-unique country. It was just too much to ask of her.

I tormented myself with the conundrum. Half of me wanted to say yes. Half of me knew that I should not. My work had trained me to be decisive, and once a decision was taken, I was able to go ahead and forget the 'might have beens.' But this decision took me days, if not weeks of agonising, talking it through over and over with my wife, who was unbelievably supportive, telling me again and again to go to Japan if I wanted to, that we'd work it out somehow. But finally, I decided, not without regret, that I simply could not.

My next trip to Japan came around quickly. I made arrangements to go see Nakamura-san, knowing that he was probably the key person behind this offer. I made my way to his office – now up in the the elevated air of the 9th floor, where the senior executives were based in NTC. Yes, I had discovered that the Nissan 'single status' myth was not totally true – the 9th floor had private offices, deep carpets, and a small army of silently efficient administration assistants to bring never-ending supplies of green tea and biscuits to visitors such as myself from inferior floors. Nakamura-san received me with his usual calm politeness – flask of tea, fan, pencil case neatly laid out as ever, arms crossed calmly, looking kindly at me through those wire-rimmed spectacles. I took a deep breath and explained to him that I could not come to Japan to take up the job offer. Extreme reaction from Nakamura-san – he gently raised both eyebrows about a half-centimetre to express his shock. Oh. Why, Dave-san? I respected him too much to try to spin him some corporate line. I told him the truth – that I did not think I had the drive for it, that I was worried about my personal life and marriage, and that I could not reasonably ask it of my wife. Don't forget that in Nakamura-san, I was talking to someone who had dedicated his whole life to the company since his early twenties – no questions, no compromises: utter dedication. I felt like some weak, soft western dilettante standing in front of a Zen master, admitting that, well, this whole Zen thing was kinda hard.

Nakamura-san thought for a while, and studied the immaculate surface of his desk. Finally he sighed and told me he understood. He'd also been thinking about it, and realised that this was a problem the company would eventually face. It was inevitable now that they would start to get Western engineers like me – people hired straight from college in Europe or the US ten or more years ago – who were now senior enough to be asked to take on 'corporate' level roles. He knew that this was a tough ask for non-Japanese folks. It was a problem the company had to think about ... and solve. It was not my fault. He understood. He was not angry with me. I felt the relief wash over me – I looked up so much to this man, that hearing his 'forgiveness' was like getting absolution from the Zen master. I sat back and felt okay about my decision ... the guilt I had been wrestling with for weeks was lifted off my shoulders.

Nakamura-san played with his fan for a bit – a sign of deep concentration. He then looked up at me shrewdly and said "Okay, so what should we do with you? You don't want to work in Production or Sales and Marketing, do you? There are many jobs in Nissan Europe in those areas, you know." I shook my head. No need to explain – he knew me well. Engineering cars was my life. I had limited interest in actually building them, and even less in selling them.

"Hmmm. I thought so. Okay, another idea. NTCE is small. Now you have grown up – you are too big for it. But you know there is another big Engineering centre in Europe. Not far away. Just the other side of the English Channel, in fact."

I sat up. He obviously meant Renault's gargantuan Technocentre, just outside Paris, the Renault equivalent of where we sat in NTC – another huge engineering centre, with upwards of 11,000 people based there. He was suggesting that maybe I could work there. This was amazing. Nakamura-san was THE Nissan guy of all Nissan guys. Here, he was suggesting that I go work for our internal rivals within the Alliance ... to leave the company, to put it baldly. But I was intrigued. Was it even possible? Of course, there were regular temporary transfers between the companies – Nissan staff who would be on a foreign assignment to Renault for between a year and three years, or French staff who would travel in the opposite direction for a similar temporary gig – but this was not what we were talking about: we were talking about a permanent move, potentially for the rest of my career. Nakamura-san promised me that he would speak to the right people in HR – in Nissan and in Renault – and in Renault Engineering, to see what we could do. I thanked him, stood up, shook his hand and made my way back down to the packed rows and much scruffier carpets of the engineering engine room.

Nakamura-san was as good as his word. It took a few months of discussion and wrangling, as we were breaking new ground for the Alliance, but eventually all parties involved accepted that I would be transferred to Renault Engineering. This would be wrapped up in the camouflage of a three-year assignment. I would be 'detached' from Nissan for a three-year mission to Renault's Technocentre, but the unwritten agreement was that – assuming that I did not do anything too stupid, or screw up too badly – at the end of those three years, I would officially resign from Nissan, and transfer job contracts to join Renault. This gave both parties some level of insurance – if Renault did not like me, it could get rid of me, back to where I had come from. If Nissan decided that after all, it could not survive without me, it still had an umbilical cord by which to yank me back to the mothership.

I travelled to the Technocentre – a short and easy business trip compared to the 12-hour flight to Japan – to be interviewed by the then head of Renault Engineering, Jacques Lacambre, who wisely wished to confirm that he was not buying a total dud. The interview went well – thankfully he spoke excellent English, so I did not (yet) have to struggle with my schoolboy French, and he struck me as a very shrewd operator indeed. The interview must have gone reasonably well, as a week or so later I received the official confirmation from Nissan HR – I was to report for duty to the Renault Technocentre in the New Year of 2006. My time at Nissan was almost over.

In the last weeks I reviewed the P32L project that I would now be handing over to my replacement, a NTCE veteran called Ian Shepherd: a very safe pair of hands indeed. It was in pretty good shape. The car was not – in hindsight – a particularly technically innovative one, apart from the basic 'crossover' concept itself – and even that was not completely new, as we've seen. To be honest, it was an intelligent interpretation of ideas culled from previous vehicles, blended together and presented to the market at the right time, very carefully executed and aggressively cost-engineered.

But it had not been without some technical challenges. I was very proud of the full-length 'panoramic' fixed glass roof, for example. Now commonplace, this was only the second such full-length glass roof that had made it to mass production in a European model (the honour of being the first was the 2002 Peugeot 307 SW 'break' – an estate or station wagon). This had been a struggle – the more conservative elements of Nissan engineering did not believe that such a large glass panel would have the strength to meet safety standards, in side crash and in roll-over in particular. Heavy pressure was put on me to keep the B-pillar roof crossbeam in place as an insurance policy. This is the strong 'roof bow' or pressed steel rail that runs between the central (or 'B') pillars between the front and rear side doors, helping to give the whole structure strength and stiffness. Normally this beam can sit there happily, hidden between the metal of the outer roof skin and the headliner. Even in cars with 'normal,' short, sunroofs, it can be neatly packaged just behind the sunroof, minding its own business and doing its discreet structural work. But of course, retaining it would defeat the whole point of a full-length panoramic glass roof. The airy impression of being open to the stars or moon would be ruined by a 120mm (5in) wide chunk of steel interrupting the view. Peugeot had made it work – and as far as I was concerned, there was no way that its engineers could do anything that we could not! The NTCE team did its homework – designing the roof section and B-pillars correctly to compensate for the 'lost' crossmember, simulating crash, NVH (Noise, Vibration and Harshness) carefully, studying the crucial adhesive technology required to keep this large glass panel in place securely even as the lift generated by high-speed air running over the convex roof surface would do its best to suck it off the car. We ran many thousands of hours of structural and aerodynamics simulations – both of which had advanced enormously since the days when six of us would share one PC. I was confident of the results, and defended them ferociously against the doubters – regardless of their rank. It duly made it to production, and was and still is a very popular customer option.

Another side-project that I was particularly proud of was the Qashqai +2. This was the seven-seat version of the car, with two small 'jump' seats squeezed in as a third row of seating. This was the brainchild of one of the Nissan Europe marketing staff, turned into reality by our packaging wizard, Tim Dunn. Tim and his team cooked up a recipe where we could – just – cram in two small folding seats in the boot/trunk area of the vehicle, with a very modest stretch in the wheelbase and a slight raise of the rear section of the roof[20]. The genius of his layout proposal was that the car was basically unchanged forward of the B-pillar, vastly reducing the investment required to launch what would be essentially a new car, addressing a part of the market then dominated by much larger 'people-carriers' or large estate cars. We immediately ran into opposition. The extra seats were too small, we were told. We could never manage this project in parallel to the base model. Anyway, NTCE was not allowed to modify the platform[21], so it was against the rules to do it. Once again, I decided that asking for forgiveness was easier than asking for permission, and went ahead regardless.

We developed detailed 2D and 3D layouts, and built a rough-and-ready wood-and-foam ergonomic 'buck' to show that our idea was indeed feasible. By now, things had changed at the very top in Japan. Patrick Pélata was back in Renault, and had been replaced as effective Number Two at Nissan by Carlos Tavares – yes, the very same Mr Tavares who had helped us so much by allowing me access to the parts cost database for the Renault Mégane. I had the buck shipped to Japan and took the considerable risk of presenting it to Mr Tavares himself. He walked around it, glanced over the layout drawing I had brought with me, and – of course – asked a few questions about the project cost estimates. He folded himself briefly into one of the two little folding jump seats while I explained the modifications from the base vehicle, before re-emerging to keep to his schedule of endless meetings. "Good idea. Do it." was the conclusion – a nod, a quick handshake and he was gone. The resulting 7-seater, or 'Qashqai+2' as it would be dubbed, would be a huge hit with young families – a limited increase in the vehicle overall size (and price) allowed parents to cram the kids' friends into the rear row of seats for short journeys to and from school, sports events or a trip to the cinema.

The original project, the base five-seater car, was also in good shape. The

---

[20] *The wheelbase was stretched by 135mm (5.3in) and the whole vehicle length by 211mm (8.3in). The rear 'lip' of the roof was raised by 38mm (1.5in).*
[21] *Modifying the wheelbase was technically a 'platform' change and hence should have been managed by Nissan Japan, in theory. But back then, I felt that rules are meant to be broken …*

Design work with Stéphane and the Design studio was essentially finished, all the Production releases were completed, and NMUK was well on its way to spinning up ready for the first Production Trials. The costs needed constant watching, like a pot of milk on a hot stove, but they were at least well understood and documented in painstaking detail – I handed that data over to Ian like a proud father might hand over his infant daughter to the baby-sitter. The project was far from finished – the teams still had more than a year of hard work awaiting them to bring the car to its successful Start of Production in January 2007. But we had at least laid solid foundations, and I was handing it over reasonably 'clean' – with few, if any, skeletons in the closets.

On the last day before the Christmas break at the end of 2005, I walked out of the door of NETC Cranfield for the last time as a Nissan employee, and took a final walk around the grounds. It had been my home, school, university and automotive church for the last 13 years. The simple white building loomed pale in the darkness, the ornamental lake was just an oil-black mirror in the night. But I knew that the Koi carp were still in there, having grown a lot bigger now than the little goldfish I had seen surfacing to be hand-fed crumbs by people coming back the staff restaurant when I had first walked around the lake as a nervous job interviewee, more than a decade earlier. I wished them well, and went on my way.

CHAPTER 8

Electric Vehicles? Seriously?

Jan 3rd, 2006. The first working day after the Christmas and New Year holiday. I was sitting nervously in the enormous visitors' reception area in the Renault Technocentre, waiting for my new boss' administration assistant to pick me up and lead me into the bowels of this enormous building. I was no longer wearing my familiar Nissan Blues armour/uniform. I was wearing a suit and tie and had a '*Visiteur*' badge pinned on my lapel, just to remind me that I was well out of my comfort zone. I listened to the musicality of spoken French around me and watched the rather exotic and glamorous denizens of this new world wander back and forth. I felt a twinge of misgiving. What the hell was I doing here? My world was back in the UK. Or maybe in Japan. But not here.

The Technocentre was an imposing place – not designed to put jumpy visitors at ease. It had been built in the early '90s – bringing together various Renault engineering and other functions, which had previously been spread out all over Paris and its environs, as Renault grew and evolved organically from its founding in 1898. Renault had been fully nationalised by the French state after WW2, as the French government took ownership of the war-torn wreckage of the then family-owned company, in circumstances that are still controversial to the present day. The French state's stake in Renault had steadily declined from 100% in the postwar years to a mere 15% at the time of the Alliance with Nissan, but the hand of the state was still felt very firmly on the tiller. Thanks to governmental aid, Renault had been able to buy a large slice of land just outside the historic town of Versailles, about 16km (10 miles) from the Arc de Triomphe, at very favourable rates. The company had erected an ultra-modern, state-of-the-art engineering centre there. Very different from the 1980s university-campus layout of NTC, with its various high-rise tower blocks dotted around Hon-Atsugi's steep green hills, Renault's Technocentre was low-rise and sprawling, with a very 1990s concrete, glass and water visual aesthetic. It was always impressive to first-time visitors: one could take them for a stroll through the central atrium of the main building called '*La Ruche*' (the Beehive) and casually point out the amenities of this engineering city – shops, a hairdresser, a bank, a dry-cleaner, and, of course, a whole selection of restaurants and cafés. This *was* France, after all. It was bewildering in its sheer size and complexity – even years later,

I would regularly get semi-lost in the depths of the building, looking for one of the thousands of meeting rooms. Which was not a problem – one could simply stop and ask a random person passing by or at their desk *"Excusez-moi de vous déranger, mais je suis ou, exactement?"* ("Sorry to disturb you, but where am I, exactly?"). As we all got lost, regularly, this appeal to our fellow colleagues' charity was usually well received, with the appropriate directions and a wry smile.

My new job was as a Section Manager in what Renault called Platform Architecture. To be honest, this came as somewhat of a shock to me. First, Section Manager was several steps down in 'rank' for me. In Nissan I was a Director. Section Manager was (roughly) a demotion of three steps in terms of seniority – back to where I had been in Nissan five years before. It rapidly became very clear to me that whatever experience I had at Nissan counted for little here. Despite seven years of the Alliance – or possibly *because* of seven years of the Alliance – whatever I had done in Nissan was of very little interest to my new colleagues. In fact, once I had swallowed my slightly-bruised pride, this made a lot of sense. One of the foundation stones of the Alliance was taking great care to keep the Nissan and Renault brands separate – which meant, to some extent, keeping the engineering teams, with their very distinct and proud histories and technical cultures, separate. I would have to work my way up through the ranks again. Fair enough – and in fact, it would take me six years to work my way back up to a director-level position again.

There were other cultural shocks. Renault is a deeply French company, and remains wedded to a great tradition of French engineering – a deeply intellectual, academic tradition. The French Revolution is not merely distant memories in dusty history books. The French pride themselves in their national meritocracy – that anyone, from whatever background, can make their way to the very top, through effort and more precisely, through academic ability. And academic ability is largely judged on one's mathematical ability. Those who survive the intensely competitive secondary or high school selection systems make it into the best university – the fabled 'École Polytechnique,' known by the very cool short-hand code of simply 'X.' Those who make it through that stage go to the ultra-prestigious 'École des Mines' (literally 'School of Mines' – these institutions still show their roots as military engineering schools, founded in the years immediately after *La Révolution*). The graduates of this system – the 1% of the 1% who can put 'X-Mines' on their CVs, are basically guaranteed a place at the very top of the French industrial pyramid – on the boards of recognised blue-chip companies like Michelin and Renault. Renault's senior management was peppered with 'X-Mines,' including Mr Ghosn and Patrick Pélata. Below 'X' there is a list of second-grade *écoles* – not quite as prestigious as the Polytechnique, but still comparable to Cambridge, Oxford or Harvard. Hence, where you went to university mattered – really mattered.

It was therefore logical – to Renault, at least – that I was asked to produce a paper copy of my own Engineering degree, to prove that I was in fact an engineer, and to judge where my Engineering school fit into the French system of university rankings. I was nevertheless somewhat taken aback – I was 35 at this stage, not 22, and nobody at Nissan has ever even asked to see my degree certificate. In fact, I had to phone my mother back in Ireland, who had the Latin-encrusted piece of faux-parchment tucked away somewhere. She kindly got it photocopied in her local Post Office and sent it off to me. I admit that this rankled a bit at the time – but I had spent the last decade in a very different environment. In Nissan's rather unique blend of Japanese tradition with Anglo-Saxon pragmatism, paper qualifications mattered little. The only thing that mattered was what you could *do*. I felt that I had already proved what I could do, and did not need the additional evidence of some scroll with a few lines of cod Latin. But I was in a new world now, and I would have to adapt to it, not the other way around. So I swallowed my pride, politely supplied the copy of my certificate to the Renault HR team, who duly fed it into the computer to see where 'University College Cork' fitted into the prestige-ranking algorithm of French Universities. It seems that it passed muster.

There were more challenges. Jacques Lacambre, the gentleman who had interviewed me before I joined Renault, was indeed as shrewd as I had suspected. He had asked me how much I knew about fundamental powertrain and platform packaging. I had been honest with him, and explained that although I had done a little bit of both during my time in Japan, I was more at home with Upper Body topics, and especially with whole-vehicle project management. I politely suggested that maybe I could take up a role as ACVE – not necessarily on a whole new project, as I was a new boy in Renault, but maybe an easier project like a small vehicle facelift program? *Bonne ideé*, said M Lacambre. He then totally ignored my suggestion and assigned me to powertrain and platform packaging. I was appalled. "David, please don't worry," he said, patting me on the shoulder in an avuncular manner. (I was instantly '*David*,' pronounced 'Dav-eed' in France. No-one in France can pronounce 'Dave,' and even if they could, it's never used as it immediately conjures up cultural memories of a particularly cheesy 1970s singer with blow-dried hair and dubious fashion sense. Best avoided.) "In platform architecture, you will build your *réseau* – what is it in English, again? – yes, yes, your network. The network is very, very important in Renault, David. Very important. In the architecture team, you will meet everyone – the designers, the engine people, the chassis people, everyone! Remember – the network! Very important. Please trust me on this." I did, and he was right – it was all-important.

Let me try to explain it like this. At Nissan, when I handed over my business card, it would say something like:

> David Twohig
> Director, Program Office
> Assistant Chief Vehicle Engineer

while at Renault, my card read:

> David Twohig
> Chef de Section,
> Ingénierie Plateforme Amont*

[*Section leader, Advanced Platform Packaging]

Now, in Nissan, when you handed the card over (with two hands, accompanied by an appropriately deep bow), a colleague would look at the job title first. The title placed you in the pyramid of the organization. Immediately, the reader would know your role, your position in the hierarchy, what you could and could not do. You could get down to business immediately, formalities over – the person knew who you were. It took me a while to realise that when I handed over my newly-printed Renault business cards to my new French colleagues, they could not give a merde for the job-title. They looked at the first line – the line that said 'David Twohig.' They wanted to know – who is this guy? Who do I know that knows him? What have I heard about him? Maybe even – what university did he go to? The title was a mere mechanical job role – in Renault, it was the person that mattered. Which is what Jacques Lacambre meant when he talked about the importance of personal networks – I knew no-one, and no-one knew me. They did not care what roles I held before – certainly not what I had done for Nissan in the UK or Japan – they needed to get to know me for themselves. And there was no way to establish this rapport except the hard way – proving yourself on the ground, and getting to know people, personally and face-to-face: cultivating the famous réseau or network. And this would take me years to do.

Next challenge: the language. My French was close to non-existent – some schoolboy French that I had learned before I dropped the subject when I was 15 or so: a big mistake, with 20:20 hindsight. Andy Palmer had actually had the foresight to encourage me to take some French evening classes when the Alliance had been signed – he rightly suspected that the language would be important in the future. But my French was rudimentary at best. I was now responsible for a team of about 14 people (much reduced from the hundreds of people I had led on the P32L project) who apparently spoke no English at all. I was to later find out that several of them actually spoke excellent English, but my first direct boss at Renault had, rather cruelly, ordered them on pain of death *not* to speak English to me – on the logic that I needed to learn French more than they needed to improve their English. So it was very close to total linguistic immersion for me – from the first day, I spoke only French, had to figure out the intricacies of a French-language AZERTY keyboard, and deal with the daily flood of emails and various documents in the language of Molière rather than that of Shakespeare. It was tough – really, really tough. For the first six months, it was possibly the hardest challenge of my career so far. I was in the typical beginner's language-learner position of translating everything in my head – everything that came in through my ears, and everything that came out of my mouth. It was exhausting – not only mentally, but physically.

I would come home to our company-rented house in a southern Paris suburb and collapse on the couch, too exhausted to do anything but eat and watch television – in English. I was so exhausted that I actually started to think that I was ill, that I had some disease. This lasted months. Until one day, the famous mental 'click' happened. I was in another interminable meeting, with maybe 10 or 15 colleagues, doing that typical Latin thing of speaking over each other at high speed, when I realised that I was following the conversation – not translating it, but simply understanding it. My brain was no longer trying to 'convert' or parse the phrases, it was simply understanding them. A wave of relief washed over me – I was getting it at last. Over the next three months or so, I became what I guess you could call fluent, and it started to become a pleasure to speak and write the language, rather than a worrisome burden. I grew to love it, and I speak it now as well as anyone who took to it in their fourth decade of life might be expected to. The next two projects that this book will discuss were both executed in French – so I will try to translate the different approach and 'flavour' of working in French rather than the Japanese-English that had been my linguistic tool so far.

The final hurdle to conquer was the actual work itself. Renault divided the Engineering team into two camps – what it called '*Métiers*' and 'projects.' 'Projects' are easy to understand – they were the vehicle-dedicated single-focus project teams like the one I had already run on B32A or P32L. '*Métier*' is a beautiful French word, much used in Renault, which is almost impossible to translate, so I hope the reader will forgive me if I use it 'as is' on occasion. If you look it up in a reputable French-English dictionary, you will find it translated as 'profession.' But that is like translating '*baguette*' as 'French bread' – it does not even begin to do justice to the noun. *Métier* is a rich blend of profession, know-how, skill and pride in craftmanship. In Renault, it denoted those Engineering teams that supported the project teams – the expert, dedicated teams in their functions, be that body engineering, electrical, powertrain, trim, whatever – like the Exteriors team I had led in Nissan before the automotive Fates took me to Japan. In Renault, the perceived wisdom was that it was not desirable to spend too much time in 'project' teams – the theory being that one might lose touch with the core expertise of one's profession, and that it was hence necessary to alternate between projects and *Métiers* to have a fully rounded career. It's actually not a bad theory, and I was to see some of the benefits later. But for now, I had been dropped into a *Métier* that was new to me, working in a language that I did not speak, and without the familiar structure of a project to cling to. Easy, right?

But it was too late to turn back, so I rolled up my sleeves and got into it. The language started to come naturally, as I've explained. And the task of my little team started to absorb me. Our task was what some car companies call 'packaging,' some call 'layout,' but Renault called 'Architecture.' This last word is actually the one I prefer, as it most accurately describes the job – very much like what an architect does in designing a new building. Specifically, my team was responsible for designing the platforms for new vehicles, and positioning their powertrains – the engines, gearboxes and various drivetrain parts – into those platforms correctly. This really is the very first task carried out on a new vehicle – the platform architects are the ones that lay the foundations, to carry on with the building analogy. They dig the footings, lay the electric cables, gas and water pipes, and decide where the walls and doors will be. Other teams will put the roof on, and much later we will need to decorate it and install kitchens and bathrooms, but the first steps are made by the architects. I gradually learned more about this science (with the usual hint of craft) and my new team were thankfully as patient in listening to me mangle the subjunctive tense of French verbs as they were in educating me on the right way to position a powertrain unit in the engine bay of an all-new platform. And Jacques Lacambre proved to be quite correct – the platform architecture teams really do interact with almost every function in a car company, from the production engineers to the Design studio, from purchasing to the materials specialists, and of course with every possible and imaginable engineering function or *Métier*. Slowly, I built up the indispensable network, and I started to be known as that ex-Nissan Irish guy who spoke French with a terrible accent.

Out team worked on many vehicles – pretty much every passenger vehicle that carries a Renault, Samsung, or Dacia[22] badge built between 2006 and about 2018 – a list too long to include here. After nine months or so, I was promoted to *Chef de Service* or General Manager of the entire platform

---

[22] *Renault's success in the '90s was such that it actually had cash left over after its investment in Nissan. Its 1999 shopping spree also including buying the almost-moribund state-owned Romanian car-maker Dacia, and a year later, a 70% stake in the automotive division of the enormous South Korean industrial conglomerate, Samsung.*

architecture team, somewhere about 60 people. It was varied, interesting work. We established a standard engine-bay architecture for many of the cars that the Renault Groupe built in those years. I interacted with my recently ex-Nissan colleagues quite a lot, working to try to share components, architectural principles and whole platforms between Nissan and Renault – never an easy task. We weathered the 2008 financial crash – as devastating an event in the automotive industry as it was in so many others. I had to 'fire' dozens of people for the first time in my career – while not Renault direct employees, as they were mainly sub-contractors, they were still part of my team, sitting in our offices with us, sharing lunch, coffees, hard times and good times. It was heart-breaking to have to walk around the offices, take them one-by-one into a small meeting room, and tell them to pack their things. But it was an experience shared by many thousands of people in those years as the financial crash wreaked havoc in daily lives.

And in those first few years in France, I grew to know more about this new company and about my newly adopted country. The two are intertwined – Renault is *deeply* French, in the same way that Nissan is deeply Japanese, and I cannot think of either company without thinking of their countries of origin. Renault was a fascinating, complex, infuriating, loveable mess. You have already heard me refer to Nissan as an army, or to use semi-military terms or analogies when I write about it. Renault was like a family, not an army. It had the weaknesses of family – disorganisation, rivalries, pettiness, everybody knowing far too much about everyone else, cliques and tantrums. But it also had the positive aspects of a family – it was a truly human company. It really *did* care about its employees, invested heavily in HR and personnel support, and took things like career development and employee welfare very seriously indeed. Where Nissan could be cold, clinical and efficient, Renault was warm, friendly, and messy. I would grow to love both, understand both and (eventually) be able to operate within both. It's something that not many people succeeded in doing, especially in Engineering – the two companies were simply too different, and people trained in one culture found it difficult to transition to the other. But I managed – at least to some extent – to do so, and I ended up working at Nissan for 13 years, and another 13 years for Renault: the perfect balance?

My world was to be shaken up – again – by a telephone call from my boss in mid-2007. It was a week or two before the all-important French summer holidays in August, and you could already feel the pre-vacation relaxation settle on the Technocentre, as people looked forward to their month-long (unthinkable in Nissan) break to go to visit the sea, the mountains, or their families in the French provinces. My then boss, a gentlemen called Philippe Guérin-Boutaud, called me to request me to come over to his office, lost in the labyrinthine depths of the Technocentre. Important, please come immediately. I threaded my way through the maze, curiosity piqued. Philippe asked me to shut the door and said "David, orders from the Boss." He then added *"M Le Président"*, just in case I had not understood that he meant *le* Big Boss. By this time Carlos Ghosn had taken over the leadership of the whole Alliance – not just Nissan, but also Renault, having replaced Louis Schweitzer as CEO of Renault in 2005. But remarkably, he had not let go the reins of Nissan. He was now CEO of *both* companies – the first, and (at the time of writing, still the only) person to lead two Fortune 500 companies at the same time. To help him try to tame Renault, he had also recalled his faithful lieutenant, Patrick Pélata, from Japan to run Renault day-to-day as COO, leaving Carlos Tavares in charge of affairs at Nissan. Ghosn and Pélata were very much calling the shots at Renault now, as they had at Nissan since the first days of the Alliance. Philippe went on to explain to me that Mr Ghosn had tasked us – in great secrecy – to start to study several EVs. EVs, as in Electric Vehicles? *Oui*, confirmed Philippe. Very confidentially, he told me that Mr Ghosn was in serious discussions with an Israeli start-up called Better Place, and that we were to study how to get an electric motor and a battery big enough to power a car into a Renault platform – quickly. I was flabbergasted. Of course, both Nissan and Renault had some history of electric vehicles ... like most car-makers, both companies had dabbled with battery-powered vehicles, in Nissan's case, at least as far back as the 1940s[23]. Nissan had recently started to work on what was to become the Nissan Leaf, to be launched in 2010, but this was not really front and centre of the stage in 2007.

Despite these various attempts over the years, if you had asked the average Nissan or Renault engineer in 2007 how important EVs would be for the companies' future, they probably would have laughed and cracked the usual stock EV jokes about golf carts, milk floats, etc. Now it was becoming clear that they were about to be elevated from back-room engineering experiments to Priority One. The Boss wanted them – and ours was not to reason why, ours was to make it happen. I went back to my desk, scraped up a skeleton crew – literally three engineers – swore them to secrecy, and briefed them with just enough information to allow them to get to work. One of them volunteered to work over the summer vacation and to start to rough out the initial designs for the first Renault EVs to be commercialised in serious volumes.

Very quickly my team sketched out a platform layout for the vehicle that was to become the Renault Fluence EV. This was, to be honest, a very flawed design, designed with two almost-insoluble constraints. First, it was a quick-and-dirty project, to the point that it went from just dirty to filthy. Better Place, the full-steam-ahead Israeli start-up that had succeeded in striking a deal with Mr Ghosn, wanted the vehicle ready yesterday. So it had

---

[23] *Nissan's involvement with EVs goes back to 1947, when the company built the adorably-cute 'Tama' all-electric utility vehicle.*

to be a cut-and-shut of an existing vehicle – the 'Fluence' sedan or saloon-car version of the Mégane that was already popular in Better Place's country of origin and first market, Israel. Second, it had to have a removeable battery. Better Place's business model was pinned to the principle of what they called battery 'Quick Drop' – the idea that you could get around the problem of slow EV recharge times by physically swapping out the battery, as you might switch the Duracell AAA cells in your TV remote control. A seductive, but in reality, deeply flawed idea, it was unfortunately a 'must' for our new partner and now semi-client, Better Place. Hence the EV version of the Fluence was designed around an e-motor up front under the hood, replacing the petrol engine of the normal car, but had a large lithium-ion battery pack, mounted upright behind the rear axle, in an orientation that allowed it to be 'dropped' vertically out of the car, in the space behind the rear axle. A quick, certainly, but not a particularly good design – this layout basically killed the trunk space, filling it instead with batteries. The balance of the car was also horrible – there was so much mass behind the rear axle that it handled like an early Porsche 911 – just not in a good way. But as I say, our hands were tied, by the ill-fated and short-lived liaison with Better Place.

Meanwhile, the engineers in my team had also started to cook up a much more interesting dish than the fast food Fluence-EV. Mr Ghosn had bet the farm on EVs – an incredible risk at the time – and would later announce that the Alliance would invest 4 billon US dollars in the technology. It is hard now to remember just how revolutionary this was in 2007-2008, when we have all become used to EVs humming around our streets alongside their fossil-fuel-burning brethren. Ghosn's decision was truly ahead of its time ... partially catalysed by the Better Place deal, we would pin the company's accelerator pedal to the carpet on EVs. Our friends in Nissan were already well advanced on the project that was to become the Leaf, the world's first affordable EV. Renault, meanwhile, was going for broke – we would launch studies on a full range of electric vehicles – four[24] in all.

My team broke into feverish activity. We dropped many of the 'traditional' projects we were working on, to start to detail the initial platform layouts for these new vehicles. But one of them in particular caught my interest. It was intended to be an electric city car – like our own Clio, Nissan's Micra or the Ford Fiesta. A 100% electric, dedicated vehicle – no hedging bets with hybrid powertrains or half-committing by package-protecting for a petrol or diesel engine, just in case we got cold feet. We had that rarest of things in the modern automotive industry – a near-blank sheet of paper. How could we best approach the design of a small car, fully optimised to be electric, not a rushed, compromised make-over like the Fluence? And most intriguingly, the Product Planning folks had set a very aggressive pricing point. This car was not

---

[24] As well as the Fluence-EV for Better Place, Renault would launch the tiny urban two-seater Twizy, the Kangoo-ZE all-electric van, and the Renault ZOE, all between 2010 and 2013.

planned as an EV for the rich eco-warriors that would soon be buying Tesla Roadsters and the first Model S. It was for everybody – a car cheap enough that the average French, German or British small city-car owner could afford it – an EV for every-(wo)man. *That* got my interest. Designing cars with no economic constraints is relatively easy. Designing cars to near-impossible cost targets – that's hard, and I now had considerable experience in doing just that.

I probably spent more time on the layout of this new small-car platform that I should have, to the detriment of the other projects, if I am being honest. I got absorbed by the details of the fundamental technical choices and layout. I was there, with two of my best engineers, when we made the first, hand-drawn outlines on a whiteboard, doing the engineering equivalents of 'jamming' on whether we'd put the e-motor up front or in the rear (we went front-mounted ...). And as the months went on, and the car's underpinnings started to take shape on the team's CAD tubes, I started to fall in love with this project. I could feel that I was getting slightly bored with leading a *Métier* team. It had been three years now, and I felt the 'project itch' – the desire to have a car that was 'mine' again – to control everything to do with it, and to make it a reality. And so I started a rather unsubtle self-promotion campaign. I basically annoyed and badgered every senior person that had the bad luck to cross my path in those months. I wheedled, whined, complained, postured and begged to get this project. I can be persistent when I put my mind to so being – and I'm convinced I probably just bored the company into submission. I asked so often for the position of Assistant (or 'Deputy' in Renault-speak) Chief Vehicle Engineer that the-powers-that-be simply caved in. They gave a figurative Gallic shrug, said *"Okay David, si tu insistes, elle est à toi, alors..."* ("Okay David, if you insist, it's all yours..."). I was back in the saddle. I had my next obsession, and it would last for the next four years.

CHAPTER 9

Uncharted waters

I started my new job as Deputy Chief Vehicle Engineer of the all-new small EV in January 2009. I was delighted to be back to 'owning' a whole-vehicle project. My new boss, the CVE, was Jean-François Simon, a Renault 'lifer' who'd had a long career in Renault's Engineering and Program Management teams. He was famous in Renault for being parsimonious with the company's and the customers' money. I settled into my new role and started to get to know my new team. As on P32L, I would indirectly manage hundreds of engineers, but my direct team would be surprisingly small – about a dozen 'core' senior engineers, to whom Renault gave the horribly clunky title of 'Technical Synthesis Engineers.' As before, I would have one of these senior, experienced people for each of the key *Métiers* or engineering functions – body, trim, electrical, chassis – and my own old team, packaging. And, of course, there was now a small team from the emerging specialist engineering disciplines of high-voltage batteries and electric powertrains. Add to that a few people to help me manage the project costs and scheduling, and we had our core engineering team, our Band of Brothers and Sisters[25] that would be tasked with engineering this very new vehicle over the next few years.

We also had a name for it. Renault loves its alpha-numeric codes as much as any other car company, and has the usual Byzantian-complex method for coding its vehicles. Our new baby was code-named B10. It would later be known as the Renault ZOE, but to us it would always and will always be B10, as Qashqai will always be P32L to me.

It's very hard now to portray just how novel it was to be designing a mass-production electric vehicle in 2009. Of course, there was already increasing social conscience about the effects of greenhouse gases and a dawning realisation that society in general and industry in particular had to 'do something.' Governments had started to talk seriously about restricting greenhouse gases such as CO_2. But we were very far from today's general acceptance of EVs as part of the solution towards slowing global warning. Tesla was still very much a hip Californian start-up. Its only product was the Roadster, which was simply a Lotus Elise body-tub, crammed with batteries

---

[25] *Renault was a much less 'masculine' company than Nissan in these days. It still had the usual engineering, and particularly automotive-engineering massive preponderance of male engineers, but at least we had some female colleagues.*

bought from Panasonic and retrofitted with Tesla's own motors. It was a curiosity, and no more. Even when the Tesla S was launched in 2010, it was far from a major event – it was a very expensive toy for Silicon Valley tech staff to spend their bonuses on, not yet seen as a serious threat to the global car industry. Mitsubishi had launched its i-Miev electric car in 2009, rebadged by Citroën as the C-Ion, but unfortunately it had a distinct whiff of the golf cart – it was small, too tall, and rather cheap and rickety. Its range was a barely-adequate 150km, and it seemed to confirm that EVs were not to be taken seriously rather than anything else.

If we had paid attention to the competition in 2009, it would not have reassured us that we were onto the next big thing in automotive history. Most major car companies were sniffy at best, openly poking fun at us at worst. Several giants of the car industry who now loudly beat the EV drum at major automotive industry events pointed the finger at the Alliance and at Mr Ghosn, and ridiculed us for the very idea of wasting money on battery-powered vehicles, of all things. Nissan had a head-start on us, having started work on the future Leaf before Mr Ghosn signed the Better Place deal that had kick-started our efforts at Renault.

Of course, I had maintained friendly contacts with my Nissan colleagues, and was therefore able to benefit from regular discussions with Kadota-san, the CVE of the Leaf project. He was to be a great help to me in the next few years – like slip-streaming racing cyclists or racing cars, I was able to 'draft' behind him, as his project was about 18 months in advance of mine. I would regularly chat to him, and he would always kindly alert me to pitfalls or bumps in the road ahead to avoid. This was to prove invaluable. Nobody else had ever tried to design truly mass-production, genuinely affordable EVs before. Having a colleague clearing the path for me, and willing to share his hard-won experience, was invaluable. The original Nissan Leaf and the first-generation Renault ZOE share very few physical components for various strategic, commercial and organisational reasons too complex and, frankly, boring to bother the reader with here, but the cars are truly linked by goodwill and friendship. Several of the most successful features on the ZOE are there because of Kadota-san's good advice, and his being able to warn me to avoid certain issues discovered by his team on the Leaf – ironically, sometimes too late to correct them on the Leaf, but with enough time to build those lessons into its little French sister.

As I became absorbed by the project, it became clear that this was a very different challenge to P32L. The original Qashqai was in some ways a very conventional vehicle. The whole-vehicle 'crossover' concept was innovative, to some extent at least, and we had gone to great pains to get the customer targeting very accurate, but the vehicle itself was quite conventional – a five-door bodyshell, quite conventional running gear, existing petrol and diesel engines from the Alliance catalogue of powertrains. Of course, it did not feel 'easy' in any way, as it was our first project and the cost targets were genuinely very hard to achieve, but from a pure engineering point of view, it was relatively simple. B10 was a whole other world. Yes, the body structure was conventional – pressed steel, similar to its Renault Clio stable-mate. Front and rear axles were also fairly conventional – a MacPherson strut up front, with a 'twist-beam' rear axle, like millions of other small hatchbacks. But everything else was new. Obviously, the electric powertrain was all new – the first-generation e-motor was developed with Continental[26], and an all-new battery had to be designed and developed from scratch by the new electric powertrain design team. But almost every other component of the vehicle also had to be rethought. One illustration of this is the number of patents filed during the development. A typical, standard new vehicle project might generate a handful of patents – 10 or a dozen would be impressive. The B10 project would generate over 70 new patents: and that is not counting those registered by the various suppliers.

So, what's so hard about designing an EV? There is a perception that it must be easier than designing a piston-engined vehicle. Surely EVs are simpler? For example, some people will tell you that the motor in an EV has just one moving part, compared to the hundreds of moving parts in an internal-combustion engine. Surely it's just a case of buying battery cells from someone, connecting them up to this simple motor and we're good to go, right? Wrong. Especially wrong when trying to design the car to a budget suitable for 'normal' folks' pockets ...

I can assure you that designing and developing an EV is just as hard as developing an ICE[27]-engined car. It certainly was in 2009 – and it still is, to a large extent. Yes, some items are simpler. There is no gearbox, for example. No fuel tank, no exhaust. But that's about it. Everything else is present, and needs to be designed and tested, without the benefits of over a century of technical precedent to call on, as we have for ICE vehicle technology. Sourcing components is a lot harder. The auto industry supply chain has developed and matured around ICE vehicles for over a hundred years – sourcing parts for EVs is still a lot harder. In 2009 it was extremely difficult.

The powertrain itself presents huge problems. Don't believe the myths about electric vehicle motors containing only 'one moving part' – that is simply untrue. The 'one rotating part,' or the rotor of the motor, is itself a very complex component, being composed of hundreds of delicate laminations, combined with very precise windings in the case of B10's 'wound rotor' motor, or with very carefully-positioned magnets in the case of other EVs like the Nissan Leaf or Tesla's vehicles. The power electronics 'driving' the

---

[26] *This bought-out e-motor would be replaced shortly after launch with an in-house Renault-built motor.*
[27] *Internal Combustion Engine, the petrol or diesel-fuelled 'piston' engines that had dominated the last century of the automobile.*

motors are highly sophisticated, combining high-speed computing with large power transistors capable of switching the currents required to control the motor in a very accurate and smooth manner. On B10, we had chosen to mount the motor in front, in a similar location to the conventional piston engines that have powered most small front-wheel drive cars since Alec Issigonis' classic 1959 Mini design. Careful attention has to be paid to cooling the motor – another myth is that EVs don't need large air intakes to cool them. Believe me, they absolutely do. Although the electric motor might not vibrate like a conventional ICE as it pulses with its own explosions, it *does* emit a high-pitched electro-magnetic whine – so careful attention must be paid to motor mountings and balancing to prevent this becoming annoying to the occupants of an otherwise almost-silent car.

We spent a lot of time on the thermal systems of the ZOE. EVs are complex things in this regard. The motors need to be cooled, as do some of the power electronics. The battery needs to be cooled most of the time – but sometimes it needs to be heated up: batteries don't like cold weather, and can't deliver their full output current when they have been parked up in a Scandinavian airport for a month while their owners were on a winter-sun vacation. And, of course, the human beings inside the car sometimes want to be cooled, sometimes need to be warmed. And here's the tricky bit ... there is no source of 'free' heat. A modern ICE is a marvellous thing, but it's still – even 140 years after its invention by Nicolaus Otto – pretty thermally inefficient. Only about 35-45% of the energy that goes into it from the petrol that it gulps is converted to kinetic or 'moving' energy. Most of the rest is wasted as heat – blown out into the air by the car's radiator and exhaust systems. But that 'waste' heat is highly convenient for engineers designing systems to heat up the chilly occupants on a cold winter's morning – they can simply duct some of the free energy into the cabin heater, and everyone is nice and toasty in no time.

There are no such freebies for the engineers trying to design an electric car. Those amazing little e-motors are 95-97% efficient – only a paltry 3-5% of the electrical energy flowing into them is wasted as heat or noise. Not enough to even warm a cigarette lighter, never mind the whole interior of the car with its shivering occupants. But my team – led by the excellent interior engineering leader, Gilles Hermer, came up with a truly brilliant proposal – the heat pump. We couldn't claim to have invented the technology – it had been used to heat and cool buildings for many decades. General Motors also applied it to the ill-fated EV1, the grand-daddy of all modern EVs, in the mid-1990s. But the technology had lain dormant since then, judged to be too complex and too expensive to apply to a mass production car. I will not attempt a technically-accurate description of how a heat pump works here

(continues on page 113)

Picture gallery

Aerial photos of Nissan Motor Manufacturing UK (NMUK) taken in 1986 (L) and in 2015 (R),
showing the impressive expansion of one of Britain's most successful car plants.
(Courtesy NMUK)

Nissan Technical Centre, Europe (NTCE), Cranfield, UK. Where the author learned his trade.
(Courtesy Nissan UK)

Early P32L (Qashqai) concept sketch, 2003. The designers are clearly searching for the 'crossover' shape. (Courtesy Nissan UK)

Right: First Qashqai concept car shown to the public at the 2004 Geneva Motor Show. The shape is getting closer to reality. (Courtesy Nissan UK)

The final production form of Qashqai emerges from the clay milling machine in Nissan Design Europe's London Paddington studio, 2005. (Courtesy Nissan Design Europe)

The 'Model Freeze' clay model – Qashqai's shape is fixed. Now we've just got to finish engineering and build it! (Courtesy Nissan Design Europe)

Prototype Qashqai hot weather testing in Spain, summer 2006. It's probably 40+°C. Note the improvised hard camouflage around the rear tailgate – the car was still top secret at this stage. (Courtesy Nissan Technical Centre Europe)

Prototype Qashqai cold weather testing in Norway, winter 2005-2006, in the dim light of the almost-24-hour Northern winter darkness. (Courtesy Nissan Technical Centre Europe)

A rightly proud Stéphane Schwarz, the Chief Designer of Qashqai, leaning on the wing of his creation. (Courtesy Nissan UK)

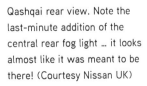

Qashqai rear view. Note the last-minute addition of the central rear fog light ... it looks almost like it was meant to be there! (Courtesy Nissan UK)

The finished object. Qashqai is launched, February 2007 – and is an immediate hit.
(Courtesy Nissan UK)

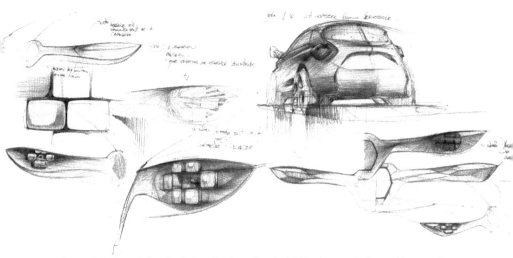

Some of the beautiful early design sketches of project B10 – the car that would become the Renault ZOE. (Courtesy Renault Design)

The 'little mouse' – Jean Smeriva's sketch that inspired the friendly front-end design of the ZOE. One of the big challenges for the engineers was to hide the charge port under the Renault 'lozenge' logo. (Courtesy Renault Design/Jean Smeriva)

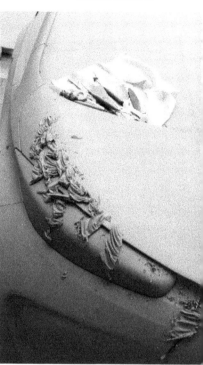

Jean Smeriva, the lead exterior designer (L) and Geoff Gardiner, Design Director (R), review some early ZOE design proposals. (Courtesy Renault Design)

Right: The shape of the final ZOE emerges from the clay… the craft of the clay modeller is still a vital one in the modern car industry. (Courtesy Renault Design)

Almost there. This is the final shape of the ZOE preview that would be shown at the 2010 Paris Motor Show to 'tease' the final production model, still two years away. (Courtesy Renault Design)

Opposite, top: The first Production Trials at the Renault Flins plant, late 2011. The first pre-production cars go down the line – high stakes.

Opposite, bottom: The very first ZOE approaches the automated 'battery marriage' station. Fingers crossed! (Both pictures courtesy Renault/Antoine La Rocca)

Prototype ZOE undergoes winter testing, Kiruna, Sweden, March 2012. Batteries don't like the cold – so the battery heating systems need to be rigorously tested.
(Courtesy Renault/Philippe Stroppa)

The final production shape of the Renault ZOE.
(Courtesy Renault)

Early Alpine sketch. How do you revive an automotive icon without resorting to cheap pastiche? The Alpine design team gave a master class in how it should be done. (Courtesy Alpine)

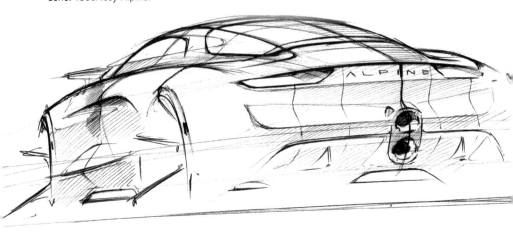

Alpine prototypes in Spain during road and track testing in 2016. The various vinyl wrap camouflages identify different iterations of prototypes. These cars would have a very hard life – and would end in the crusher. (Courtesy Dan Prosser)

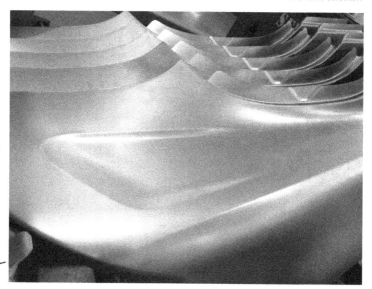

The oh-so-troublesome A110 rear wing/fender pressings – lots of blood, sweat and tears would be shed to get these right, and hence realise the design team's vision of a spoilerless sports car. (Author's collection)

Winter testing in Sweden, 2016. L-R: the author, Thierry Annequin, Alpine chassis wizard, and Bernard Ollivier, the Boss. (Courtesy Lapland Ice Driving/Felix Macias)

Engineering cars is not all stress and long days in the office. The little Alpine dances as well on a frozen lake as it does on a Spanish race circuit. (Courtesy Lapland Ice Driving/Felix Macias)

Geneva Motor Show, March 2017. Antony Villain (Alpine's Design Director) and Bernard Ollivier deservedly get the honour of revealing the car that would end Alpine enthusiasts' 30-year wait. (Courtesy Alpine/Olivier Martin-Gambier)

The image that still makes the author cringe. The *Top Gear* press car burning at the side of a road in the French Alps, 29th January 2018. (Courtesy Iain May)

This is what's left when a car that's 97% aluminium burns. Not a lot – but enough to piece together the events that led to the fire. (Author's collection)

The Alpine A110. Most definitely worth all the effort. (Courtesy Alpine/Yannick Brossard)

– the hardcore engineering geeks among the readers will no doubt look it up – but suffice it to say that it's a gadget that reverses the normal operation of an air-conditioning system. It actually captures the energy in the ambient air surrounding the vehicle – and there is a lot of energy available, even at low ambient temperatures – and 'pumps' it (as the name suggests) into the cabin to heat the occupants. It's possible to design a heat pump with what engineers call a 3:1 coefficient of performance – ie, for every 1kW that you spend in driving the pump's compressor, you 'capture' 3kW – hence gaining 2kW for free. Sounds like black magic or perpetual motion, right? It's not quite that, but it's damned clever. Gilles and his backroom boys in the interiors and thermal systems teams managed it, designing a practical, reliable heat pump that was the first to be installed on a mass-production[28] car.

It's not an exaggeration to say that my team and I were obsessed by safety – terrified by the prospect of making an error and injuring or, even worse, killing a customer. Of course, this is something all Chief Engineers of high-volume production cars must learn to live with – knowing, statistically, that someone will be severely injured or will die in one of the cars that we build. But we were very conscious that this was a new type of car – if there were to be any injuries or (almost unthinkable) deaths due to a preventable error of design, we would never be able to look ourselves in the mirror again. Hence, we were even more conscious than usual of the risk. Renault and the Alliance were taking a huge bet here – in the teeth of public mockery. We were investing billions in a technology that had been ignored for decades, with only Elon Musk and his merry team in California playing alongside us. If we were to have any safety issues, we would kill the electric car: again, and this time, forever.

Hence, I spent countless hours with the safety engineers analysing all possible crash scenarios – frontal, rear, roll-over, oblique, at all possible combinations of angle, speed and temperature. The tricky one, we quickly realised, was the so-called pole impact – a chilling test that simulates the car losing control, and spinning/sliding into an immovable cylindrical object like a tree, lamp-post or pole. In testing, we do the opposite – we fire the pole into the side of the car at high speed to simulate the crash. This is a tough test for a 'normal' car – the body structure, doors, seats, seatbelts and side airbags (if fitted) have to be very carefully designed to protect the fragile human occupants from the enormous forces generated by these impacts. But in B10 we faced a new challenge – the floor of the car was in fact a giant sandwich, four layers of metal filled with the battery itself. To get enough autonomy or driving range, we packed every tiny cranny with

---

[28] *With all due respect to the GM EV1 – and I am a huge admirer – only 1117 were built: which hardly qualifies it as mass-production.*

battery cells, filling the space lengthwise between the two axles, and the full width of the floor almost up to the sills. Hence the problem with pole impact – the pole would (if the car's structure was not well designed ...) chop into the battery like an axe, damaging the cells and releasing the flammable liquid electrolyte inside them, with disastrous and fiery consequences. A chilling thought, especially as it is possible that the occupants in such a crash may not be in a fit state to get out of the vehicle: doors may be jammed, or the occupants may be unconscious, or simply too stunned to extricate themselves. We thought long and hard about how to prevent such an event, and how to help the emergency services respond, should the worst come to the worst.

Many days, weeks and, eventually, months were spent refining the tiniest design details of the body structure, the battery housing, and the internal layout of the battery to ensure that this would never happen. Digital simulations would be played through step-by-step, like advancing a movie frame-by-frame in the old days, carefully watching how the metals and plastics deformed, straining our eyes for the smallest hint of a material reaching its mechanical limits. These digital simulations were then checked by physical tests of partial mock-ups of the car, pneumatic rams firing poles into prototypes again and again – and yet again – until we were satisfied with the results. We probably over-designed the car, and it was no surprise – but still a great satisfaction – to us when it was later to be awarded the highest crash safety rating by the independent safety watchdog EuroNCAP when it was launched.

Similar attention was paid to electrical safety. The electrical power backbone of the car ran at a system voltage of 400V – almost twice the mains electricity voltage in most European countries, close to four times as high as the 110V that is standard in the US or Japan. Lethal, if touched by a human being. All of the 'high-tension' cables were specially insulated and armoured, coloured bright orange to visually identify them from the normal, 12V cables in the car that are perfectly safe to touch. But what happens in a crash? Surely the cables might be torn, contact the metal parts of the car and hence electrocute the occupants or those that arrive to help them? The answer is no – this cannot happen. But only because the engineers put in the time to make sure it does not.

Our team designed in multiple layers of safety – some of them still trade secrets, so I will not speak in detail about them. But some are now known technology, as competitors have torn down, analysed and copied the first mass-market EVs. The first layer of safety is an easy, obvious one – link the high voltage system to the airbag deployment circuits. If the car crashes, the airbag system uses accelerometers to detect the crash and deploy or 'bang out' the airbags. It's pretty easy to arrange for the airbag unit to send a signal to the battery management system to tell it to cut off the 400V juice at the same time. This is done by opening large contactors or relays – a bit like the big circuit-breakers in your home's fuse box that might occasionally trip in an electrical storm, or when you plug in one appliance too many. But that alone is not enough. You see, the 'Throw the switch, Igor!' command cannot arrive quite quickly enough. The 'g' sensors for the airbag systems are themselves almost instantaneous. But the airbag control unit needs a little time to 'think' and decide what do so with the input from these sensors. The encrypted and redundant (meaning doubled or backed-up) signals from the airbag control unit to the Battery Management System (or BMS) brain take a few milliseconds to travel down the metre or so of copper wire separating them. The BMS also needs a short time to check the input signal and make sure that this is not a false alert. Finally, the power relay within the battery itself is a big, chunky piece of copper and iron – it cannot simply flick open in an instant, like a transistor can. Long story short, it may take 50 to 70 milliseconds to turn off the high-voltage power in the event of a crash. Of course, 70ms is not even a blink of a human eye – that's about 300ms, so it's about a quarter of a blink of an eye. But it's quite long enough to hurt you if you were to be exposed to 400V. So it is crucial to control exactly what happens to those thick orange cables carrying the power in that critical quarter-blink of an eye.

Again, our team stared at computer monitors simulating crashes, watching the action frame-by-frame, millisecond by millisecond, as the panels twisted, components crumpled and the front or rear of the car folded up like delicate origami in slow-motion. We positioned, re-positioned, protected, moulded and teased the routings of these oh-so-important orange cables with an attention to detail bordering on obsession. Until finally we were satisfied.

There were many other challenges, too long and complex to go into in any detail – I am sure I would lose the interest of all but the hardcore automotive-engineering fans were I to do so. Some were obvious technical challenges to overcome, some were extreme 'corner' or 'edge cases' – like the scenario dreamed up by one of the battery engineers, who came up with a nightmare scenario after the terrible flooding events of February 2010, where unusually high tides coupled with storm Xynthia flooded low-lying areas of the Vendée department in the south-west of France, leading to the tragic loss of dozens of lives. The poor engineer, probably having watched this disaster on the evening news, woke up with this unlikely chain of events in mind: a ZOE owner parks their car in a well-sealed garage for the night. The garage then partially floods with sea-water (it has to be sea-water, as the salt content makes it slightly more conductive …). If the water rose to *just* the right level in the garage – not too high, because that would fully short out the battery – the voltage inside the battery could theoretically electrolyse the water, splitting the H_2O into H_2 (hydrogen) and O_2 (oxygen).

If the garage was very well-sealed indeed, the H_2 just *might* be able to accumulate in sufficient quantities to become explosive. If our incredibly unlucky ZOE customer then had the misfortune to come down to their partially-flooded garage to investigate, and continued their streak of bad fortune by also having a faulty light-switch that sparked (or, in true Gothic horror movie-victim fashion, they decided for some reason to carry around a naked flame like a candle) the garage could, in theory, go 'Boom!' So yes, the little ZOE also has countermeasures built in to avoid even such far-flung scenarios.

As Qashqai had done, the car slowly made its way off the drawings, CAD tubes and workstations running endless safety simulations and into physical being in the hundreds of components required. Again, the first vehicles were platform 'mules' – this time they were bastardised Renault Modus donor cars, ugly, dumpy little primer-grey and black testbeds, running silently around the test track in Aubevoye in the sleepy Norman countryside to the west of the Technocentre: Renault's equivalent of Nissan's Tochigi proving ground back in Japan. I had my first extensive experience of driving EVs in these vehicles, testing and comparing them to the very few EV competitors that we could then get our hands on back then – including of course, back-to-back comparisons with B10's Japanese sister, specially-imported pre-production examples of the Nissan Leaf.

Just as for the Qashqai project, I had to count the pennies, or rather euro-cents, very carefully. It was always key to me that this car was not an elitist 'eco-car' – it was in the grand tradition of cars like the Citroën 2CV, the VW Beetle and Renault's own 4CV – cars that had been designed to be affordable to everybody, and to pull a bomb-ravaged Europe out of the economic devastation left by WW2. Our car was aimed at providing affordable, zero-emissions transport for people of average financial means, ultimately with the aim of being no more expensive to run than a 'normal' small family car. This was a very tall order at the time – EV-specific parts like the electric motors, power electronics, and of course the battery itself were still considerably more expensive than the equivalent systems in conventional cars. To have a hope of making the whole vehicle come out at a price point somewhere near its conventional piston-engined rivals meant iron cost discipline on the rest of the car – the body structure, chassis parts, interior and exterior trim. Luckily by now I had almost a decade of experience in aggressive cost-management – from the bad old survival days of the NRP 3-3-3 as Nissan fought its way out of the grave, to the endless rounds of cost-optimisation of the P32L project. Scraping cents out of the cost of conventional parts of B10 in order to re-invest them in the new EV-specific systems was almost second nature to me now ...

The engineering challenges of the new technology were tough, but I could feel that we had enormous support from the company. Carlos Ghosn's bet on EV was a very public gamble. He had re-stated the Alliance's ambition to become the global leader in zero-emissions vehicles at various Press events and in countless interviews. Our PR department had already started to 'leak' snippets about the new project – a teaser of the car called the Renault 'ZOE[29] Concept' was shown publicly as early as September 2009 at the Frankfurt Motor Show. In 2010 this was followed up by the 'ZOE Preview' at the Paris Motor Show, a 90% accurate version of the final vehicle that was still more than two years away. These carefully-co-ordinated PR efforts built up expectation in the international Press, watching carefully to see whether Carlos Ghosn's huge bet would pay out. As a result, the machinery of the Alliance swung slowly but powerfully in line, to back up the Boss' declarations. EVs were now Priority One, and I felt the power of the giant company behind us.

Engineering efforts are a constant fight for resources – financial, physical (for things like expensive test facilities such as wind-tunnels or test rigs) and, crucially, human – for the best engineers and technicians. Projects sometimes feel like riding a bicycle: some feel easy, like riding along with a nice tailwind, pedals spinning easily as the miles roll by; others are like fighting uphill into a howling headwind, shoulders hunched over the handlebars, thighs burning. In terms of availability of resources, B10 was the former – resources were always available for the asking, support of the various other departments just a 'phone call or an e-mail away. The project was technically extremely difficult, and the cost targets were stretched beyond belief, but we always felt that reassuring tailwind that came from the unequivocal and unanimous support of the company's boardroom.

Working on projects this intense often feels like living and working in a bubble. You come to work, concentrate for hours on the car, working with (largely) the same people, day after day. Days turn into weeks. Weekends are just a pit-stop. Months are replaced with project milestones. Years and seasons become demarcated by the winter testing for ABS and summer testing of cooling systems. It's a very inwards-looking world ... a kind of automotive submarine voyage. As our team lived in this artificially insulated world, strange things were happening in the company. Minor issues like the vaguely-farcical legal battle over the correct way to write the name of the car. But other, far more serious, legal battles broke out in 2009 and 2010. The so-called 'fake Chinese spying affair' threatened Renault, the leadership of Carlos Ghosn and the Alliance itself. This bizarre affair was very much

---

[29] *The name 'ZOE' had been registered by Renault as early as 1991. It means 'energy' or 'life' in archaic Greek. It would later be the subject of a lawsuit deposed by several French women called Zoé Renault (Renault being a reasonably common family name in France), who sued Renault for using a version of their name without permission. The case was thrown out by the judge, but it was decided to capitalise the name of the car and to avoid the use of the accent 'aigu' to avoid any possible confusion – hence, ZOE, and not Zoé. I stick to the correct form in this text to avoid annoying Mmes Z Renault, or their lawyers, any further ...*

centred on the intense development efforts around EVs. It is a long, sordid and complicated story, but the gist of it is that three managers in the company – one of whom worked directly on the EV Program team – were accused of stealing industrial secrets on EVs (including B10) and of selling them to 'China.' They were summarily sacked, amid a deafening media storm in France and beyond. It would eventually turn out to be a hoax – a scam set up by a member of the Renault in-house security service to extort money from the company. The three accused were totally innocent, and ended up being exonerated and financially compensated for their unfair dismissal. But the 'affair' damaged the company significantly – and cost it one of its most brilliant leaders, Patrick Pélata. Patrick felt he needed to resign in the wake of the scandal, and did indeed leave the company. Carlos Ghosn's hitherto-impeccable record as CEO of Renault – and Nissan – was also injured: he was particularly criticised for two uncharacteristically ill-judged media appearances on French television evening news. But he survived, and the company plugged on, sadly without Patrick – replaced as COO of Renault by another gentleman we've already met several times – Carlos Tavares, recalled from Nissan in Japan to his *alma mater*. These momentous events did not quite pass us by, immersed in our B10 submarine-bubble as we were, but almost.

You will know the next steps now from the P32L project – finalising the interior and exterior design with the Design Director, Geoff Gardiner, and his team, selecting the various suppliers, approving the detailed engineering specifications of each part, large and small. As the months rolled on, we completed these steps, and were finally ready to build the first 'real' B10s. Like the P32L prototype we built in Nissan's Zama prototype build shop, the first B10s would be built in Renault's impressive prototype build facility, part of the giant Technocentre complex. For several weeks my team camped out at the prototype mini-factory, as the first cars that actually looked like what we had created in the Design Studio took shape. Of course, we encountered all the usual problems of first builds – parts that did not quite fit first time, and the inevitable missing parts – sometimes our fault, for having released the parts too late, or having made a badly-judged late change; sometimes the fault of the various suppliers, struggling to produce the very first components with new tools, processes and staff. But slowly we solved each of the problems, and the first 'ZOEs' were parked up in the confidential storage areas, carefully camouflaged and safe from prying eyes or low-flying aircraft.

Mr Ghosn himself even dropped in to see the vehicles, and requested to test drive one, including a very short test on public roads. This was always stressful in a pure prototype – the chances of it actually breaking down were not negligible, and lots of precautions had to be taken – cars in front to clear the route, cars behind with bodyguards ready to sweep in to pluck the CEO out of any embarrassing or potentially dangerous situations. We took this very seriously – a previous Renault CEO, George Besse, had died violently, assassinated on his Paris doorstep in 1986 by extreme-leftist militants. Carlos Ghosn was not universally liked in France, and his public profile and very recognisable physique made him a very real potential target. Testing unusual-looking and not necessarily reliable prototypes on public roads was a stressful event, and armed bodyguards were as important as a fully charged battery. Luckily, all went well, Mr Ghosn enjoyed a silent and smooth ride around the roads near the Technocentre, and safely delivered the car back to us, to undergo a lot more unpleasant torture at the hands of the test and durability engineers.

CHAPTER 10

Expectations meet reality

As 2011 rolled into 2012, the expectations and hopes vested in this new project were mounting. Maybe to levels that were simply too high. Part of this was entirely of our own making. Our Zero Emissions drive in general, and ZOE in particular, was a big deal, and the Renault press and PR machines were beating the media drums at every opportunity. Renault was struggling to maintain market share and was falling behind Nissan in terms of sales and, especially, profits.

The project was also a big deal for the Flins plant, an hour or so to the west of Paris, where the car would be built. Flins had fallen into hard times in recent years. The plant was built in 1952 as Renault moved car production out of the centre of Paris into a high-tech, purpose-built new plant on the banks of the lazy River Seine, in what was then quiet farmland. It was a typical grandiose post-WW2 industrial and social project, designed not just as a car factory, but as a whole new town, with its own schools, hospital and power plants. In its glory days in the 1970s it employed a remarkable 21,000 people and built over 400,000 vehicles a year, including such French automotive icons as the Dauphine, Renault 5, 18 and Clio. However, volumes fell and it became increasingly difficult to profitably build small cars in Western Europe: the economic pressure was relentless to build these cars in Eastern Europe or even further afield, where raw materials and labour costs were lower. Flins was now a sad old giant, with endless swathes of empty buildings, broken glass in its windows, deserted wastelands of cracked concrete parking lots and just a single production line, slowly churning out a mere fraction of what it had built in its heyday. The down-at-heel main office building had chipped lino and fly-blown photographs of VIP visits from a bygone era – a bizarre mix of Nikita Krushchev, Queen Elizabeth II, and the Queen of '60s France, Brigitte Bardot, touring black-and-white lines of half-built cars and waving to enthusiastic workers. I always suspected that the waves for BB were rather more enthusiastic than those for either Her Majesty or for the First Secretary. The decision to build ZOE here in Flins was a major coup for the plant – building the highest-tech new small car for many years, back in the industrial heartland of France, a figurative stone's throw from the Eiffel Tower.

The ZOE project was a big deal at national level – the fact that it was a

French car, high-tech, designed and *made* in France, played very well indeed with the government of the time, led by the mercurial Nicolas Sarkozy, who had very publicly complained about Carlos Ghosn 'allowing' Renault to off-shore production of cars like the Renault Clio to lower-cost countries like Turkey. It also played well in the game of Alliance-friendly (or less friendly) rivalry between Nissan and Renault. Renault was confident that its car would be cheaper, better-looking and technically more sophisticated than its Japanese sister, the Nissan Leaf. This would help restore French pride, given that Nissan had beaten us to the EV market, the Leaf being launched in late 2010, and was now consistently the stronger of the Alliance partners in terms of both sales volumes and profits. A fact that rankled with the proud French managers, who remembered only too well that it was Renault that had 'saved' Nissan, only a short decade ago. The Program and PR folks showed the project off at every opportunity, to VIPs, visiting Ministers and to the automotive press. The pressure and expectation ramped up at a national and even international level. We had even 'launched' the car – statically at least – at the Geneva Motor Show, in March 2012, to much fanfare and acclaim. Mr Ghosn had promised, very publicly, that the first cars would be delivered to customers by the end of 2012 – a promise that would come back to haunt us. As the year wore on and the technical challenges continued to mount, the months and weeks would start to run out, fast.

Another hint that things were not going to be a smooth ride was the very first production, or rather pre-production trial, at Flins. It went far too well. As I've mentioned, carmakers love their acronyms, and the first trial of a production car in the factory in which it is destined to be built is usually called something like PT1 (Production Trial 1), PS1 (Pre-Series 1) or something of that ilk. But regardless of the code/jargon, the process is pretty much the same, and it's always a key milestone towards mass production. The goal here is to build some vehicles that are 'off-tool' and 'off-process,' but not yet 'off-cadence.' What does all that mean? Well, 'off-tool' means that the vehicle's component parts should be coming out of the production tools[30] that will make them in 'real' production. In other words, the parts are no longer prototypes. The injection-moulded plastic parts like headlights or door trims must be produced by their respective suppliers in the real, geometrically accurate, tool-grade, hard-steel tools that will make the final parts for real customers. The stamped metal parts must be stamped with the final, massive stamping dies that will bash out (hopefully) millions of those panels for the production vehicle. Same for the forged, die-cast or machined mechanical parts – and vitally, also for the electronic components. These

---

[30] *The word 'tools' as used in the automotive industry does not mean hand tools used by humans like spanners (wrenches), drills, etc. It means the equipment used to fabricate the components – often press tools used to stamp out sheet-metal parts, or injection-moulding tools used to shape plastic or aluminium parts.*

latter must be made with production-intent printed-circuit boards, the final chipsets and hardware selected by the suppliers and the carmaker. Software, however, is a notoriously grey zone – the very definition of 'off-tool' shows that it's not very well adapted to software – software (sadly) does not pop out of physical tools, therefore the very definition of 'prototype' and 'final' software is much harder to define than for hardware. But the reader will get the principle – the idea is that *all* component parts for the production trial are coming out of the same industrial processes that will produce the final vehicle.

Of course, that's the target, but in reality, no carmaker achieves 100% 'off-tool' exactly as planned. Inevitably, reality intervenes, and something will run late. A supplier may have to make a late change to its part based on prototype testing results. Very often, the carmaker decides that a change has to be made – to the great annoyance of the supplier, who may have thought that their part was 'good to go.' Cost, or rather cost-*reduction* drives many a late change – there may be an opportunity to save a few cents on a part, but it will often take a few extra weeks or months to re-tool in order to realise this cost reduction, hence missing a deadline. Finally, you have the truly unexpected that always happens.

On Qashqai, we had an injection mould for a headlamp drop off a forklift as it was loaded into a truck to transport from the toolmaker to the suppliers' site. Die cracked, a million euros gone, and six months to re-machine it, just in a moment of carelessness. A fire destroyed a supplier's facility on ZOE, taking all the pre-series off-tool parts with it, forcing us to fall back on the previous, prototype, parts. On another Renault project – not one of mine, thankfully – a flood in a brake systems supplier plant in South-East Asia meant that some brake component tools were under several metres of filthy flood water. On that occasion, the company paid for industrial scuba-divers to dive into the murk and fish the tools back to the surface to keep the project moving.

So, there is always a proportion of the vehicle that simply can't be off-tool. This is one of the key jobs for the Chief Engineer. Part by part, you have to 'make the call' of whether the part really *must* be off-tool, and the build should be delayed, or whether it is acceptable to have a prototype part in its place. It's a very pure example of where trade-offs must be made between the points of the classic project-manager's 'iron triangle' of Quality, Cost and Delivery. The Quality department will explain that this is very easy – all parts must be 100% off-tool: anything else would be folly and will put the end customer at risk. The production plant boss will agree with this, nine times out of ten – if the damned design engineers have not got all the parts ready on time for the production trial, well, they must be lazy or incompetent, or both, and the Chief Engineer obviously needs to pull their finger out. Sadly, it's usually a bit more nuanced than that – the clock is ticking, and every day that the production trial is delayed, the cash-in from sales of the vehicle gets pushed back. So quality needs to be balanced with timing, or delivery. Also, that late part that's been delayed in order to re-engineer it to save 30 cents, will improve the business case. So sometimes it's actually the right decision to accept that a brake line or an armrest will not be off-tool, and that the production trial should go ahead with a less-than-100% pure off-tool vehicle. It was my job to make this call, on a part-by-part basis, explain these decisions to my colleagues from Quality and from Production, and of course be responsible for the outcome as we went ahead – if the prototype part that I had accepted on the vehicle failed a test in six months' time, caused a major headache, or (ye Automotive Gods forbid) caused an accident, that was on my head.

'Off-process' is easier to understand, and also generally less controversial. It basically means that the car will be built for the first time on the 'production line' itself. The prototypes we talked about in previous chapters were essentially hand-built – either in stationary work stations, where the car does not move, but the parts and human beings come to the car (this is how low-volume luxury or sports cars are often built even in full series production, by the way) or on mini production lines that are very different to a real factory. At PT1 the car will really, physically, go 'down the line' in the factory for the first time. The production line will use the real workstations, tools and equipment that will be used to build the car in mass production. This allows the manufacturing plant to test if everything is accessible correctly, if the production workers can easily access all fasteners, interconnections, clips etc, and if the line-side equipment to fill the various fluids like coolant, brake fluid and air-conditioning gases work properly.

Finally, 'off-cadence.' In mass production, the vehicle is built at incredible speed. Very successful high-volume products like Qashqai are built at what is known in the jargon as 50- to 60-jph or 'jigs per hour.' This is the line 'cadence' or, put simply, line speed. A 'jig' is a historical word, another of those rare survivors from the world of coachbuilding, which these days describes the cradles that hold the vehicles securely as they make their way down the production line. 60 jigs per hour means that one new car pops off the production line every minute. That means that the production workers at each workstation have about 55 seconds to complete their tasks. I cannot convey to you just how impressive this is, in reality. If the reader ever gets the opportunity to tour a car production line running at full chat, I beg of you, please do take the opportunity. The production line workers in a plant like Nissan Sunderland or Renault Flins are incredible. They do not carry out simply one task in their allotted 55 seconds. Most will fit several components, with different types of fasteners – be they clips, threaded fasteners, adhesives, or some combination thereof. Some of the parts are large, heavy, and with complex shapes, making it physically difficult to manipulate them into position. Many of the parts are fragile – they can be scratched, broken or contaminated by careless handling. And the men and women who do this job, do it with the seemingly effortless grace of professional athletes – which they are, in many ways. If I sound like an awestruck admirer of their skill, it is because I am – the two weeks I spent in my youth at Nissan trying to do their jobs as part of my graduate trainee 'line training' left an indelible impression on me for the rest of my career. It was the best way imaginable to train young, arrogant, desk-bound engineers with the ink still wet on their degree certificates, to never, ever forget how hard a job the people on the production line have. It hurt my body and my pride so much that I never have forgotten it, even 30 years later ...

At the first production trials, of course, the plant does not try to build at full speed. The line – which is, in most cases, building the current production car that is due to be replaced by the new model to be trialled – will be slowed to maybe one-fifth of normal production speed. Of course, this can only be done at lunch time, or when the line is otherwise shut down – some plants run only two shifts, so this can be done during the third shift. In busy plants running three shifts, the trial run may well have to be done over weekends or other holiday periods. Typically, five or more 'jigs' are cleared in front of and behind the new vehicle. This allows the plant to cope with disasters – if the new car gets 'stuck' in some way (yes, that can happen) then this buys the time and space to get it unstuck before having to clear the line, re-accelerate it and get back to producing the current model. So the first trial production cars built are not 'off-cadence' – they are built a lot slower than they will eventually be. The plant will – again, like a professional sports team, slowly train itself to increase the cadence via the various Production Trials (anything from two to five trials before SOP (Start of Production), depending on the carmakers' approach and the complexity of the new product).

The first Production Trial is often the opportunity for the production plant to step up and flex its muscles. Until now in the project, the production team can tend to feel a little hard-done-by – all the attention has been directed at the designers, product planners and product engineers; the manufacturing folks can feel a little left out. Added to this is the fact that they often have a certain 'chippiness' – they pride themselves (not unreasonably) as being the salt-of-the-earth, hard-working blue-collar types that are going to fix all the silly mistakes made by a bunch of soft-handed, lily-livered Design wusses with black polo-necks and expensive eyewear, aided and abetted by those bloody design engineers who never had to build a car in their lives. These are enduring stereotypes, and therefore very hard to break. To further complicate matters, the automotive production plant is an astoundingly macho environment, even to the present day. The vast majority of the workers are male, and it really is a last bastion of industrial testosterone. These are not places for the politically correct, or for those that are averse to some friendly or not-so-friendly shouting and swearing. Hence the PT1 trial is often the time when 'the Plant' will shoulder aside the wimpy designers and engineers, proudly flex its biceps and generally pass the message "Okay, this is our baby now," in a fairly unsubtle manner.

Fortunately, I had always got on reasonably well with production engineering – my early training at Nissan helped, where from Day One, respect for production and for the production line staff was drummed into us. I also tried very hard to avoid the arrogance of the design engineer. Amazingly simple things went a long way. When I visited a production line, to see exactly how a part was fitted or a process was executed close-up, in person (old Ogawa-san's *genba* training "Did you check with your own eyes?" had not been lost on me) it was a simple courtesy to find the foreman or supervisor for that section of the line, and politely ask him if it was okay if I stepped onto the line at such-and-such a work station for a moment. It was even easier to simply chat to the guys on that workstation for a moment – say hello, ask them how things were going, promise not to get in their way for too long. In Flins in particular, the guys on the line almost all came from the local industrial towns that had grown up around the plant since the 1950s. Tough, hard-working lads, who could chew up and spit out some neck-tie-wearing paper-pusher like me for breakfast. It did not pay to try to flaunt Chief Engineer stripes here – that was a quick way to get nowhere.

Despite my reasonably good relations with the plant, the very first PT1 vehicle that goes down the line is still a moment of great stress. I got up early, one misty day towards the end of 2011, to drive the hour-and-a-half from my home to Flins, to 'follow the build' as the expression goes. I could already see the unmistakable shape of the half-clad ZOE body hanging off the '*balancelles*' – the giant hook-shaped hanging jigs that transport the cars along the line. I felt once again the thrill of finally seeing that shape – until now only ever seen in Design studios, R&D labs or secret test-tracks, finally in a real production plant, albeit a plant running in production-trial slow motion. I walked the line alongside the car as it was slowly pieced together – the first car is built by the most experienced line workers, the foreman and senior technicians, so that they can 'learn' the car and hence train their team later on how to build it.

It was a slow, controlled flurry of activity – some of the team manipulating the parts and wielding the air-guns, others referring to the assembly drawings and 'op-sheets' or Operation Sheets produced by my team, designed to tell them how exactly to assemble the vehicle – in what order each part was to be assembled, with which fasteners, right down to the precise torque to be applied to each nut, bolt and screw. Think a huge Ikea assembly instruction pack, but with 8000 parts, many more special tools, and 55 seconds to do each operation. Oh, and if you get it wrong, it's not just that you will have a wonky shelf on your Billy® bookcase. Torque a brake hose wrong and someone might die. Fail to apply a wax injection properly into that sill section and there might be 10,000 cars in five years' time to recall, with a 100 million euro (or dollar) warranty bill. No pressure.

One or other of the build team would occasionally wander over to me to ask for clarification of one of the documents, or ask me to note if a part was not going on as easily as it should. But all went remarkably well that day – the only drama was when what the French call the so-called battery 'marriage' station (so much more elegant than the phrase English-speaking engineers use – the battery 'stuff' station) got stuck. This was a huge moving platform, weighing five or six tonnes, that carried all the underfloor components – such as axles, brake and coolant lines, electric motor, cooling systems, etc ... and, of course, the crucial battery, a 275kg (607lb) aluminium-and-steel-box, designed to resist crash, fire, electromagnetic radiation and any other kind of abuse that you can imagine. The battery housing needed to mate perfectly with the body itself – there was a suitable cavity engineered into the body underfloor, with four very complex latches that held the battery very firmly into place: firmly enough to resist several '*g*' of negative deceleration, in case the vehicle was ever in a violent roll-over accident. The mechanical tolerances of these latches were very fine, so the moving platform (or '*platine*' in French) needed to move very precisely upward under the body, pick up some very precise guide holes in the vehicle's sills, then guide the underfloor components – including the battery – snugly into place in exactly the right spot. Automatic pneumatic screwing heads built into the *platine* would then screw in the threaded fasteners to precisely the torque settings that the body and battery engineers in my team had calculated.

Why such a complex system? Good question. The answer is as much social as it is technical. The system is designed specifically to avoid any production line staff having to carry out 'arms in the air' operations. This was a big deal in any car plant, and doubly so in Flins. If you've ever done up a bolt on the underside of a car while it was over your head, either on a lift or while you were standing in a garage pit, you will know that it hurts after a few minutes. Same deal if you've ever had the misfortune to paint a ceiling. Lifting your arm above your shoulder for any length of time is tough work – the normal muscular fatigue of holding it up there exacerbated by the fact that your arm is now above the level of your heart – hence maintaining the blood flow to the muscles is harder. Therefore, modern car plants try to reduce 'arms-up' operations as much as possible. If you tour a car plant in Japan, South Korea, China or even the southern States of the US, you will still see plenty of operations where the operator has to lift their arms over the shoulder, but Renault was a French car-maker, and heavily influenced by the still-very-strong French trade unions (akin to labor unions in the US). Arms-in-the-air operations were an ongoing bone of contention with the unions – they claimed (with some justification) that such operations were technically necessary, and that they discriminated against older or less fit production line staff. Each and every such operation thus became a battleground in a century-old struggle between management and unions. So ZOE had been designed to avoid these battles where possible – hence the automatic air-guns built into the *platine*, which would do the work that would otherwise have to be done arms-in-the-air by a human being.

The automatic guns were pretty sophisticated things in their own right – capable of 'finding' the head to the appropriate bolt, slowly spinning it to engage the thread correctly, accelerating hard to screw it in all the way as fast as possible, then progressively slowing down to measure the precise 'click' at which the bolt/nut was precisely torqued – to an accuracy far better than the best human mechanic could achieve with a beautifully-calibrated Snap-On® torque wrench. Not only that – each air-gun was connected to the plant's computer system so that it would record the precise torque applied to every fastener, on each and every car, so that we could still find, in 10 years time, that bolt number 3 on ZOE number chassis number xyz0000123 had been torqued to 45.3Nm. Necessary data to keep in these days when Product Liability law is not to be taken lightly. So, all in all, a pretty impressive piece of kit, this battery-car marriage station.

Except that it didn't work. The body of the very first ZOE came to a stop right above the famous *platine*. The huge metal bed started to move slowly upwards under the impulsion of its shiny new hydraulic rams. The battery nestled precisely into the cavity it had been designed for and disappeared from view, as planned. The automatic air-guns whizzed, connections clicked and all seemed well. The *platine* went to disengage and drop away ... only it didn't. Pneumatics whined, hydraulic pumps cycled. The station tried again to pull away, and failed again. All hell broke loose. An orange rotating light started flashing to signal a line-stop. The ear-splitting klaxon burst into life – normally a line-stop in a plant was the close-to-worst case scenario – millions of euros being lost every minute the line was down. Thankfully someone killed the klaxon – this was, after all, a line trial, not real production. But the place was still a hive of activity. French voices were raised in Latin indignation, people in hard hats started to swarm about the station. Ruggedised laptops were flipped open and control panels opened. More and more experts swarmed around the station. I took a step back – I knew I could not help now, and just prayed that it was not a cock-up on my part. I knew the penalty if the guys found an error in the product design. If the holes on the batteries did not align correctly with the holes on the body *by design*, that was my fault and I would have to carry the can.

A stressful half-hour later, all was well. Some furious diagnostics work and stripping down of the station revealed that a locking solenoid had simply received the wrong command from the industrial computer controlling the station. It was quickly bypassed manually, and the control code would be re-jigged for the next trial. In other words, a minor production engineering snafu – some poor production engineer would doubtless get a rollicking tonight, but the good news for me was that the product (ie, the vehicle) was okay ... I could breathe a sigh of relief. At least, I could on this issue – but of course, the trial generated literally hundreds of smaller points – missing parts, parts that fitted badly, operations that were ergonomically difficult, or took too much time for the operator to fit (and often both), parts that were too fragile, etc. So my team still had plenty of homework – hundreds if not thousands of small engineering changes to be made before the next trial, all of which would need to be carefully checked and signed off, of course. But taken overall, it was a success – and possibly lulled me into a false sense of security. We might have been able to build the car pretty smoothly, but we were very far from the finish line.

CHAPTER 11

High tension

High tension, or high 'voltage,' is the voltage at which electricity can really cause damage. The numerical definition depends on context, but in the automotive industry the magic figure is 60V – above that, in theory, an electrical shock can really hurt you. The powertrain in our little ZOE was designed to work at 400V – more than enough to hurt or even kill a human being. So, most definitely high tension. As we pushed on through the various production trials, I was seriously feeling that tension, figuratively speaking – the project was very much under the top management spotlight, and I knew that the pressure would only increase as we approached Start of Production (SOP). But little did I know just how much the pressure would increase, or how difficult the last few weeks and months of the project would prove to be.

If I thought that the relatively successful first production trial in the Flins plant in meant that the rest of the project would be similarly smooth sailing, I was wrong – very wrong indeed. As the pre-production vehicles were gradually assembled in batches in Flins and shipped in closed confidential car transporters to the various Renault test sites and R&D labs that would put them through their paces, we started to run into problems. Problems with charging the battery.

The first issue was relatively benign. Some of the prototypes would start to charge when plugged into their fixed, wall-mounted chargers, but they would then 'drop out' after a while – simply stop charging and sit there. This was traced pretty quickly to a complex electronic issue within the BMS or Battery Management System – the electronic 'brain' that sits within the battery of an electric car, monitoring each of the dozens or hundreds of cells within the vehicle, constantly measuring the voltage across the terminals of each cell, their temperature and, crucially, the current flowing out of each cell (when the battery is discharging as it supplies energy to the motors), or conversely, flowing back *into* the cell as the battery is charged. The control units that monitor all of this are impressive pieces of technology, and most of their electronic brain power is dedicated to safety. The nominal voltage of the ZOE battery pack was 400V. If you touch a 'live' conductor at 400V, the current is likely to travel to earth through your feet – and the route through your arm, down your torso and into your legs passes uncomfortably close to that electrically-fired and sadly non-redundant pump that we call our heart.

That current is quite enough to instantly swamp the signals running along our nerves and stop that pump. Not good. We duly paid a lot of attention to monitoring the electrical currents flowing in and around the car.

To keep the maths simple, let's say there are 100A (amperes, the unit of electrical current) flowing into the car through the charger socket that its owner has just plugged into the charging port. And let's say the car has 96 battery cells. Now suppose that the BMS reads that each of these cells is seeing a nice healthy current in-flow of 1A. Great. That makes 96a – hang on, what the hell happened to the other 4A? These 4 missing amperes are known as 'leakage current.' Very low levels of leakage current – microamps – are okay. Electrical induction or very small parasitic resistances can allow very low levels of current to bleed to earth – not a big problem. But more than a few milliamps *could* theoretically be a big problem, because it could be leaking to earth via a person's body! A human being, to an electric vehicle's 'brain' is merely a big, wet squishy electrical resistance – and if that resistance touches 400V somehow (not impossible – we could have a broken connector, or someone might do something stupid, like cutting deliberately into one of the orange armoured 400V cables under the hood of the vehicle with a penknife ...) that resistance would cause a large leakage current as the electrons, instead of going nicely into the battery to charge it, decide to play hooky and run to earth through our customer's soft tissues. So, the BMS and the various other electronic control units in a ZOE (or any other well-designed EV) constantly monitor leakage currents, and if they go above a few milliamps, immediately shut the systems down – stopping battery charging, in this instance.

On our pre-production cars, we came across a rather exotic problem, where a signal cable within the battery was inducing an electrical current in the battery casing, and 'fooling' the BMS into thinking that it was seeing a leakage current. The battery supplier – a well-known South Korean specialist – was laudably reactive. As soon as we suspected the issue, a team of experts were despatched, and I was happy to see our powertrain engineers, electrical engineers and their Korean supplier partners leaning over a disembowelled battery, poking at the offending cables and peering industriously into oscilloscopes. Sure enough, the issue was tracked down, and (as is often the case) a dual hardware and software solution was introduced for the next batch of cars to be built. Job done ... next issue.

Sadly, our charging woes continued, however. As we moved forward with ironing out the problems detected during the production trials, we started to run into other issues, always something to do with charging. First it was the charge connector itself – contacts burning out after a few hundred cycles. No problem – I gathered together a little task force, pulled in a few favours from the best electrical experts in the company, and that issue was solved with some material changes and tightening up of the already very fine tolerances on the male and female pins of the connectors. Next issue up was a lot more sinister. Some of the heavy-duty transistors, known as MOSFETs[31], within the charging system started to burn out. Opening the charger – not simple, as it was buried deep in the entrails of the motor drive unit itself – released that characteristic burnt-caramel smell of toasted electronics, the damage bad enough to see with the naked eye, carbon-black soot blown all over the circuits. This one was much trickier to solve. As is often the case, the powertrain engineers blamed the electrical engineers, the electrical engineers claiming innocence but pointing their fingers at the 'dirty' signal coming from the charger hardware (an easy target, as it was made by external suppliers).

My job, as it was often to be, was to pull people together into one room, persuade them to try to forget WHO was at fault, and to concentrate instead on WHAT the problem was and HOW to fix it. I had by now developed a comprehensive managerial toolbox to do this – the key tools within it were charm, flattery, begging, and cajoling. But the toolbox also had a second drawer – one that I tried my best not to open too often – containing threats, cold menace, and sheer bloody aggression. It was a case of using them all to a greater or lesser extent. It was also key to stay focussed on the slow, methodical analysis steps that Nissan had drummed into me years before – the pedantic posing of the famous 'Five Why'[32] questions, and the step-by-step application of a tool known as Failure Tree Analysis or FTA[33]. Most importantly, *never* jumping to the solution. Engineers love to do that, and the more brilliant they are, the more they display this weakness – "I know what it is. Let's change such-and-such. That will *definitely* fix it!" This is a trap – no matter how clever an engineer you may be, you probably do not know what the real cause is, and even if you do, it's highly unlikely to be the only root cause. Single points of failure, or a single thing going wrong in isolation, are exceedingly rare in a system as complex as a modern car, let alone a modern EV. I was never the most brilliant engineer, but I was trained to be a very methodical one, and I had the ability to impose the grinding discipline of finding *all* the root causes, and chasing them down to the bone. No glamour, no movie-scene 'Eureka' moments, just dogged technical trench-fighting to solve a problem that our customer would never care about, or

---

[31] *The MOSFET or Metal Oxide Semiconductor Field Effect Transistor is a type of high-power transistor or 'electronic switch' that was invented in 1959, but did not become widely available industrially until the 1980s. It is one of the key enablers that allowed the fast-switching, reliable, low-cost power electronic control units required to control EV motors that would be developed in the 1990s.*

[32] *The 'Five Whys' is a technique of finding the root cause of a technical failure, not just the superficial first-level cause. It involved simply repeating the question 'Why did that happen' five times to dig down into underlying causes. Very basic, very obvious ... and amazingly powerful.*

[33] *The FTA is a visual problem-solving methodology in which a failure is broken down into its root causes, following each 'branch' of the tree right to its conclusion – an actionable correction measure. FTA and 'Five Whys' work naturally together. Done right, they are annoyingly pedantic, and very thorough.*

even hear about – because we would make damned sure that it would never see production.

Meanwhile, I pushed my team to solve the hundreds of *other* issues – the tiny black plastic finisher on the rear side window that got scratched too easily after repeated car washes, the blue chrome on the Renault "lozenge" badge that peeled a little too easily, the beautiful white seat fabrics that got dirty in the Flins plant as the production staff got into and out of the car (easy fix: bag 'em up in plastic that stayed on until the car got to the end customer – but another 0.1 euro cost increase to be offset somewhere else). The driver displays that mysteriously blacked out – at first we believed that this particular issue was caused by electromagnetic interference from French military airfields' early-warning radars, until we finally figured out that it was interference from the 900MHz (900 million cycles/second) mobile phone bandwidth. As phones move towards the edge of a cell tower signal zone, they will typically increase their output power, in a desperate attempt to 'hold' the signal – ramping up to sometimes 1 Watt of radio power or more. This was just enough power to 'knock over' the driver's display.

I was lucky enough to hit on this particular primary cause myself – driving a pre-production car home late one night, I had lazily (and dangerously) put my phone on the dash, a few inches from the driver's display. As I drove away from the Technocentre and into the winding country roads that led to my home, a 30 minutes or so drive into the beautiful countryside to the south-west of Versailles, I noticed the driver's display black out. I immediately stopped and started to think it through. I happened to know that this particular stretch of road was where you lost cell signal, as you drove out of the signal area around the Technocentre and into a dead stretch for a few kilometres, before picking up cell coverage in the next decent-sized village. Could it be so-incidence that the phone was just next to the display as it blacked out? No. There are no co-incidences in engineering. I drove on, until the phone duly captured some signal, and excitedly phoned my Electrical engineering lead. "Yann, think it's the damned phone GSM[34] band. Killing the display as the phone ramps up the power towards the edge of a cell."

"Can't be, boss. We've tested that. The supplier's shown us the test reports. It's clean."

"Yann, trust me. Get them to run it again. Get our own lads at the EMC labs[35] to bombard the bloody thing. There's something."

It turned out I was right, on this occasion. Again, a software and hardware interaction. The fix involved changing some simple electronic components

---

[34] *GSM or Global System for Mobile Communications – the official international body that allocates frequency bands for cell phones.*
[35] *Electro-Magnetic Compatibility chambers, sealed to radio waves, are used to test both the emissions of radio waves from the car, and their susceptibility to being bombarded by radio waves from the exterior.*

within the display *and* re-writing some embedded code. Multiple over-lapping root causes: multiple, interdependent fixes.

But all these things were relatively 'normal business' on any new car project. Where the pressure really started to mount was, once again, on the charging systems. Remember, this was back in 2012. EV technology was really in its infancy – although the Nissan Leaf and the Tesla S were now on sale, the basic technology was still far from mature. And Renault had decided to develop a rather unusual charging system on the ZOE rather than purchasing one off the shelf. Patented as the *Caméléon*® system, it was on paper a very clever invention. Most electric vehicles have completely separate 'slow charge' and 'fast charge' systems. Slow charge (up to, say, 11kW) was delivered by AC current – the normal alternating-current 'stuff' that comes out of the wall socket. This was fed into an AC connector on the car, and into an onboard charger, that would convert the 220V AC current to the 400V direct current (DC) required to charge the battery. Fast charging (above 11kW), meanwhile, was handled very differently. The conversion from AC to DC was done, not in the car, but off-board, in the fixed charge infrastructure. Big charging stations, almost the size of a petrol pump, would convert the AC power from the grid to 400V DC current. This was then plugged into the car using a *separate* connector to the 'slow charge' or AC connector. This was the system used by the (then very few) other EVs in the market.

But the Renault electric powertrain engineers came up with a totally different solution. The system would use the power electronics already in the car (called an inverter) to control the electric motors, 'in reverse,' to convert AC current to DC. Explaining how this works in detail would take far too long, and bore most of my dear readers to death, but suffice it to say that the large inductance of the motor's fixed portion or stator, when switched using the very same power transistors that were burning out earlier in the project, could be used to make a rectifier: a system used to convert alternating current to DC. Thus there would be no need for large, expensive off-board 'fast chargers' to do this rectification step outside the car. The car could charge in 'slow' (up to 11kW), semi-fast (22kW), and fast (43kW) modes, but always from an AC source. Hence the *'Caméléon'*® moniker that the Sales & Marketing folk dreamed up – the car itself would adapt to whatever AC source was plugged into it, like the little reptile that can change its colour to match its surroundings.

It was a very ingenious concept, and still is. From day one, the charger worked beautifully – on paper. But like many technically advanced solutions, it was finicky. It became clear that the charger was very sensitive to the 'quality' of the electricity flowing into it from the charge port. I can almost see your eyebrows raise quizzically. Electricity is electricity, is it not? Electrons are electrons, no? Well, yes and no. The electricity flowing out of your wall socket is probably 220 or 230V, depending on which country you are in (110V if you are in North America), and is theoretically delivered in a beautiful, smooth sine wave form, at a frequency of 50 or 60Hz. The voltage is relative to an 'earth' – theoretically a zero-resistance, infinitely thick wire connected to Mother Earth herself ... those green-and-yellow-striped wires in your home. In reality none of these things are true – the voltage fluctuates a little, as does the frequency. The smooth sine wave is, in fact, overlaid with myriad of noisy spikes, generated by things like large electrical motors plugged into the network. And the earth connection is almost never a zero-ohm 'pure' short-circuit to the centre of the earth – it usually has some parasitic resistance in the circuit somewhere.

Without going too much into the detail (it would be long and overly technical, and some of the technology involved is still under patent) it turned out that EV charging systems can be very sensitive, both to noise on the power wave-form and to any parasitic impedance in the ground connection. If the sine wave was 'dirty' and/or the earth connection was not as good as it might be, the car might *start* to charge, but would then 'drop out' at any moment. Annoyingly, this could happen within 5 seconds, 5 minutes, or 5 hours of the charge starting. And of course, this was a show-stopper for any EV – customers absolutely depend on the car charging reliably, especially when they plug the car in at night, relying on it to be fully charged and ready to take them to work, or to drop the kids off at school the next morning. 99% reliability was not okay here – it had to be 99.97%, or better. And the key issue was keeping the system sensitive to the famous 'leakage currents' explained above – it would be easy to simply 'blind' the charger to the various noise, spikes and ground surges, and make it super-reliable. But remember our leakage currents? If the system were blinded by simply turning down its sensitivity, it might miss those few milliamperes of leakage current that might – just might – be leaking to earth through a human body. It was a tricky balancing act between charging reliability and safety. But in fact, zero compromise could be made on the latter. We needed 99.97% charge reliability, but 100% safety. And this proved extraordinarily hard to achieve ...

Meanwhile, the public profile of the car was being inexorably raised. ZOE was one of the indisputable stars of the March 2012 Geneva Motor Show. We put some – very carefully prepared – pre-production vehicles on static display on the Renault stand, to show that final production shape was indeed, very close to the ZOE Preview that we had shown at the Paris Motor Show in September 2010. It was again a big success. Folks seemed to like the friendly 'little mouse' styling. Provisional pricing was announced – as intended, significantly undercutting the Nissan Leaf, and of course a mere fraction of the cost of the only other 'serious' EV out there – Tesla's Model S. French chests swelled with patriotic pride, and one could almost hear the pressure ratchet up another click or two. I attended the Paris show very briefly, but could take little pleasure in it – I knew we still had a very long road to go. We had a clear deadline – Carlos Ghosn announced that the first cars would be delivered by end 2012 – now just eight months away. All eyes were now even more on this project – internally, from Nissan, from our competitors (BMW had announced its own small, relatively affordable EV – the car that would become the BMW i3), from the French government, and from the world's automotive Press.

Bad news travels fast ... and tends to rise to the top. The news of our charging gremlins certainly did. As we've already seen, by this time, the 'top' at Renault was Carlos Tavares. Officially he was now COO, having replaced Patrick Pélata as Ghosn's Number Two at Renault when Patrick paid the price for the fake Chinese spying fiasco. And Mr Tavares had not changed – he was still very much the data-driven engineer that he always was (and still is, I imagine) and interested in one thing only: results. No margin for lateness, over-spending or any other laxness on his watch. So, when the ZOE project started to miss development deadlines because of our repeated charging problems, Mr Tavares swiftly demanded clear reports and recovery plans. Very quickly the team found itself in the rather uncomfortable position of reporting to him weekly, supplying detailed data on each software and hardware issue we found and solved, and exactly how that would improve charging reliability. These meetings were not exactly encouraging arm-around-the-shoulder 'you're doing a great job, guys' affairs. They were hard-nosed, data-focused meetings, with zero tolerance of any sign of bullshit, or hint of hiding bad news.

My time was now almost 100% on the charging issue, leaving very little time for the myriad other 'normal' things that need to be done. As usual in times of high stress, the team tended to fracture – the engineers of different disciplines would (understandably) try to point the finger at 'the other guy,' and of course it was always easy for us all to blame the component suppliers rather than accept those errors that were certainly ours, not theirs. My job was to try to remain calm, treat the problems like the logical, mathematical issues they were, and try to transmit this rather inhuman mindset to very human, very tired and overworked engineers. The pressure on the team mounted, the hours grew longer. I went to work before dawn and did not see my wife until late at night when I would come home, re-heat something that she'd cooked for us both, then collapse into bed and sleep like the dead, until the next ZOE Groundhog day. The powertrain and electrical folks in my team lived the same grinding stress.

One Friday night in autumn 2012, I decided to take a break and take my wife into Paris for a nice dinner. Forget ZOE for a while. Unfortunately, I had booked a restaurant plumb in the centre of Paris, just off the Champs-Elysées. We had to walk a block or two on Paris' most famous street, beautifully illuminated as always. Our route took us past *l'Atelier*, Renault's famous showroom, that's been in the same spot, Number 53 Champs-Elysées, since 1910. My wife drew my attention to a beautiful backlit display on the two-storey high plate-glass windows of the *Atelier*, at least a meter high and several metres long, so that all of Paris could see it. It read "Zero Emissions: XXX:XX:XX hours:mins:secs." As I watched, to my horror, it counted down, second by second, towards a deadline ...

"That's pretty," she said. "What's it counting down to?"

"The SOP of my damned car. Let's go." I muttered, grabbed her by the elbow and moved swiftly on. Our little ZOE was haunting my steps, it seemed.

I knew already that the countdown was, if not impossible, highly improbable. And how ironic that some of the most intractable of our charging difficulties were right here on our doorstep – in Paris, the very city where the company was confidently promising to deliver the car in the next few months. Paris had recently invested heavily in EV infrastructure, through a joint venture with a company called *Autolib*. Many of the city's streets had been dug up over the past year or two to install nearly 4000 charging points, all over the city. It was, at the time, one of the most extensive EV charging projects in the world, and much of the funding had come out of the French taxpayer's pocket. Renault was 15% owned by the same taxpayers, in the form of the French state. So, it was *highly* embarrassing that the pre-production ZOEs stubbornly refused to charge reliably when connected to these *Autolib* charge stations. They might start, charge a little then flounce off (electronically speaking) to sulk in the corner. The causes were multiple and too technical to go into in detail, but among other things, the electricity supply was 'noisy' in many places in Paris and the ground connections were highly variable in the city. But we were making – albeit slow – progress. Little by little, the engineers identified the necessary hardware modifications and software fixes to 'harden' the car, making it less susceptible to these unpredictable things, but as always, with zero compromise on system safety. We negotiated carefully with the company that installed the charge stations to make some minor modifications to 'clean up' the signals. Every week I reported steady but slow progress, in the reports that went to Mr Tavares – from 97% reliability to 98%, to 99%, then 99.1%, 99.2% etc. Given time, we would get to the better-than-99.97% target.

The progress was steady, but too slow for Mr Tavares. His patience exhausted, he had finally had enough in one of the regular reporting meetings. I was not there that day – it was my boss Marc Soulas who took the brunt of his ire. Marc was a solid veteran Renault engineer, who had by now taken over responsibility for all of Renault's EV developments – four projects in all, one of which was my baby. He quickly reported back to me what Mr Tavares had asked. Or rather, ordered: "Okay, enough of these statistics. I'm tired of your predictions and forecasts. No more theory and projections. I want you to test every charge point in Paris."

Marc has broad shoulders and was pretty cool under fire, but even he gulped at this. He knew that there were 4000-odd charging points, and that we only had a few dozen cars. He also knew that testing each one took several hours. The maths simply did not work.

"Errrm, that will be complicated – you see, there are a few thousand charge points, and the availability of prototypes is pretty limited."

"I understand that it's not easy. But that's your job. Make it happen."

End of conversation. Next day I held a council of war with the core ZOE team. This really did seem like Mission Impossible. The limiting factor was the number of cars – we only had 15 or so road-legal cars available. There was only one way to do this – the whole team would have to work round-the-clock, taking shifts driving these cars into Paris, connecting them to the *AutoLib* charge points, and recording what happened. So we did – everyone in the core team was assigned a list of charge stations, in different sectors of Paris. I did my own stint in the first few days. I booked out one of our ratty, much-abused test cars, and drove it into central Paris. I then had to take out membership of the charge company's subscription service to get a credit-card sized swipe card to 'unlock' the charge points. This was done at a kiosk with an operator on the other end of a screen. Name, license number, credit card, easy. Make and model of car? Renault, ZOE. Eyebrows raised – not on the database. Makes sense – the car is not even launched yet. I talked the operator around that hurdle somehow. Registration number? I gave the prototype 'W' trade-plate number that all these pre-production cars wore. More eyebrow-action. Who owns the car? Renault, Ltd. This time she just gave me a dead-eyed stare, decided this was too much hassle, and issued me the card without further comment.

I spent the next few days driving the car round my allotted sector of Paris, plugging it in to each charge point for a few hours, moving it a few meters to the next one, repeat, then driving a short distance to the next 'block' of charge points. And of course the cars were pre-launch – it was strictly forbidden to let them out of your sight. Luckily Paris is full of sidewalk cafés, so occasionally one could grab a coffee and *patisserie* while watching over the car as it charged. My core project team members all did the same thing, working night and day, so that we could measure every charge point in Paris. And we succeeded. We delivered the report to Mr Tavares in two weeks, as he had requested. Better still, the actual numbers matched our theoretical predictions, as near as dammit.

Remarkably, even as we churned through the tough technical issues and the sleep deficit deepened, we seemed to find ways to increase the pressure on ourselves. As we drove forward through PT2 (the second Production Trial) in Flins, with the vehicles now really starting to look like real production cars (to the uneducated eye, at least), we actually made our first sale – well before the vital Start of Production milestone. The French Minister for 'Productivity Improvement,' Arnaud Montebourg, had decided that he wanted to 'own' the very first ZOE. Young, good-looking and energetic, M Montebourg clearly liked the positive PR that would come about through his being the very first customer for this, the golden child of the revitalization of French industry and on his watch. And, of course, Renault's Press and PR folks liked the idea too. So we'd committed to selling the first car to the Ministry, to be used as 100% 'green' Ministerial transport. Great. But we'd committed to delivering the car before Christmas, in line with Mr Ghosn's previous public commitments. Less great ... the pressure in the cooker notched up another few psi.

I admit that it was sometimes hard to bear. I would be lost in the latest statistical reports for some of the more exotic charging scenarios – those in Norway, for example, where the very specific method of earthing the public electricity network in that country caused us some very thorny issues with the way the charging system coped with the ground impedances. The phone would ring and a plant Quality guy would call me to say "David, you have to get out here [to Flins] *now*, we have a huge problem."

My heart would be in my mouth – another new charging issue? What the hell? NOOOOOOO!

"Okay, what is it? I can jump in the car and be there in an hour or so."

"Great. Yeah, it's a huge issue. We are still getting scratches on the centre console finisher. Your guys have to do something!"

I felt like screaming, "Screw the centre console! I've got a government Minister on my back, waiting for his fucking car and the thing does not charge worth a damn. Just put some bloody protective film on the centre console, or tell your guys to be more damn careful!" But of course I could not. The issue *was* serious – to the plant quality guys – and they could neither know the seriousness of the charging problems, nor help to solve them in any way. I had to bite my tongue, calmly suggest that my lead interiors engineer, Gilles, would probably be able to handle it, and reassure them that he'd be jumping in *his* car to speed to the plant in the next 10 minutes. The plant still grumbled and complained that I was not spending enough time there on *their* issues, but part of the Chief Engineer's role is to call the priorities correctly. If everyone is unhappy with you, you're probably doing things roughly right. And of course, Gilles solved that issue easily, as he did dozens of others.

Eventually, of course, we got there. One by one the issues were identified, root causes isolated, and hardware and/or software fixes put in place. The charging system eventually became as reliable as conventional non-Caméléon® onboard chargers, and probably more so. We *were* able to deliver the very first 'sales' car to M *Le Ministre* Montebourg, just before Christmas

2012, and hence fulfil Mr Ghosn's promise. Truth be told, we had not yet finalised all of the software patches when the car was handed over to M Montebourg in the courtyard of the Elysée Palace, in front of a phalanx of journalists and flashing cameras. Some of the powertrain technicians *may* have paid a few surreptitious post-sales visits to the Ministry's secure garage to update the car with the final production-level software and fixes. Maybe.

Slowly, laboriously, the car came to life. The final production trials at Flins went smoothly, the size of the build batches increasing, the line cadence slowly climbing up to full production speed, each batch of cars with fewer and fewer of the minute problems that needed to be solved in parallel with the charging drama. You could now see more and more pre-production cars on the public roads around the Technocentre and the plant at Flins as we piled on the kilometres, ironing out the last tiny niggles like the rattles and squeaks that are so annoying in an otherwise-near-silent EV. I drove these cars daily myself now, using them as my normal daily transport, quickly getting used to double-takes by people who had never seen one, enjoying the privilege and pleasure of giving some close friends and family their first experience of the silence and smooth acceleration of an EV. The pressure slowly lifted as we approached and passed the all-important official Start of Production milestone, and those huge, long-deserted car parks at Flins started to fill with real production cars under their plastic film coverings, waiting for delivery to eager customers, all around France, then Europe. We were there ... five years after that first meeting where my packaging team and I sketched on a white-board, wondering where we might best put the e-motors.

Lisbon, Portugal. March 2013. The final, 'dynamic' press launch of the vehicle ... as opposed to the static display of the car at the Geneva Motor Show a year before. It felt like the end of a very long road. Our charging issues were finally behind us ... although it felt like they had cost me a part of my soul, and certainly more than a few grey hairs. But the charging issues and other minor last-minute problems had delayed this dynamic launch until almost the end of the first quarter of 2013. The automotive Press were suspicious. Motoring journalists are far from stupid. They could sense that something was up – how did Renault deliver a car to some French Minister in December, but have not allowed us, the experts, to drive the car until now? But finally, the Press launch was here – always a *grand moment*, as the French say. Many hundreds of millions of euros have been spent. Thousands of people – not to mention the few hundred in my own team – had laboured for up to five years on this thing. Blood, sweat, passion and tears had been shed. Marriages had ended because of the stress. Kids had not seen their *mamas* or *papas* for months on end, as they put in the long hours needed to give birth to a product as complex as this. And now a bunch of car journalists would pass judgement, based on a couple of hours behind the wheel. This was the first time I had attended a full-on Press launch – I had already left

Nissan when the equivalent Qashqai event had been held. I was slightly stressed, but only slightly– I knew the final car was good, and I was confident that it would hold up to any scrutiny.

I flew in with a charter-plane load of journalists from Paris for the first few days of the event, to be held in and around Cascais, just to the west of Lisbon: a spot with reliable weather, a very nice hotel, great roads running up and down the west coast, and plenty of scenic spots for the photographers. I was not expected to attend the entire three-week event, as I could not afford that amount of time away from the project – there were still a million small teething problems to sort out as the production ramped up – but as the 'chief geek,' I was expected to be there for the three most influential countries – France (of course), Germany, and the UK. The event was fantastic – three days of driving the Press cars on the beautiful roads that our team had carefully reconnoitred, chatting with the journalists when they stopped for coffee and photo-sessions, then fielding more questions over dinner. The same dinner every night, by the way, and the same questions, more or less. The trick at Press launches is to answer the same question: "So, what does this heat pump really do? Why did you decide to put that in the car?" with the same energy and sincerity on the tenth occasion as you did on the first day when the first journalist asked you. Look them in the eyes, smile, don't be an arrogant swine, and do not even *attempt* to trot out any incomprehensible engineer-jargon to impress them: it really won't. They've seen all of this before, many times. If you can add a joke or make them laugh with your answer, great.

The Press reaction was excellent – not off-the-scale excellent, but very solid. Even the hard-bitten home-grown French Press, well known for their cynicism about French products, were pretty kind. A few made comments about *me* – nothing nasty, but a few remarks along the lines of 'Renault imported a Nissan guy to help then design their new electric car. Don't we have any decent French engineers anymore?' which made me laugh, being so far from the truth. But overall, the event was a huge stress relief – as I checked out of the hotel after all the journalists had gone on the last day, I picked up one of the Press cars, parked all alone in the courtyard of the Cascais hotel, with a note on the windscreen "David, please take this car back to the airport." I jumped in, and drove this wonderfully quiet and smooth vehicle that 'I' (and a few hundred others) had engineered through the pretty streets of Lisbon to the airport, listening to music and watching the Atlantic surf roll in on the beaches to the west of the city centre. I could feel the enormous pressure of the last couple of years fall off my shoulders. We had done it. Despite the technical challenges, in-fighting, politics and pressure, the car was here. It was beautiful (well, in my eyes it was). It worked. Real people could afford it – not just the well-heeled. And the world's Press had basically just patted us on the back. I was a happy man. Mission accomplished.

CHAPTER 12

Anti-climax

I've previously described the atmosphere of high-pressure vehicle projects as being akin to working in a bubble. And when the vehicle is finally launched, well, that bubble bursts. The excitement and satisfaction of the intense Start of Production, Press Launch and Start of Sales period is soon over ... leaving something of a void. Hours and days weigh heavily, enlivened only by the inevitable, and frankly welcome, teething problems to solve as the production volumes ramp up, the suppliers sometimes struggle to cope, production processes are perfected, and the first customers discover any small mistakes that might have slipped through the development net.

I was never a huge fan of being involved in what carmakers call 'life cycle management' – in other words, looking after cars already in production. Not because it is easy or boring – it's neither of those things. Besides fixing the quality snafus mentioned above, there is endless pressure to cost-optimise the product – planning and executing endless minor modifications to make the vehicles cheaper to build, and hence more profitable. It's not unusual for a vehicle to be unprofitable at Start of Production, only to start to make money one or two years into its five- to eight-year lifetime, as the engineers and purchasing specialists slowly find more efficient ways to build what is basically the same product. Any downtime between improving quality and reducing costs is filled with managing the vehicle – traditionally, this means updating the vehicle with one or more face-lifts to keep it fresh in the market and to help it compete with more recently arrived competitors: these might be more-or-less major front- and rear-end design exercises, new powertrains to keep up with emissions regulations, new interior features or, increasingly, software updates to keep the car fresh. This work is varied, interesting, and vital to the economic health of a car company – it's just not for me. I simply could not get the same kick out of managing and optimising an existing vehicle than I did out of the more creative process of conjuring a new vehicle out of thin air. Hence, in the early spring of 2013, I started to look for a new job, post-ZOE.

And one new job in particular had more than caught my eye. We've already crossed paths more than once with Carlos Tavares, now COO of Renault. Mr Tavares was a genuine 'car guy,' to use that threadbare old phrase, much beloved of those that see the need to claim such things. He spent his weekends racing cars on his own time, and with funds from his own pocket. And like all Renault 'car guys,' he had a soft spot for Alpine. But unlike most fans of this almost-mythical French brand, he had the ear of the CEO of an industry giant, destined to become the world's largest carmaker in a few years' time. Carlos Tavares had persuaded Carlos Ghosn to resurrect Alpine ...

I will keep the necessary resume of the Alpine brand history short. Alpine is a strange little brand – it inspires passion, nostalgia, very strong opinions – and oceans of ink. One could fill a very respectable bookcase with books and articles about Alpine. Admittedly, most of those are in French, so I will fill in the blanks for the non-Francophones or anyone who is not familiar with the story. I invite the hardcore Alpine fans to leaf forward a page or two!

Alpine was founded by a French gentleman called Jean Rédélé – the three *accents aigus* mean that you hit each 'e' hard in pronunciation: something like 'Ruh-duh-lay.' Born in 1922, he was the son of a comfortably bourgeois Parisian family that had made their fortune in pre-WW2 car dealerships as well as in other businesses. Jean grew up car-mad, and, more specifically, motorsport-mad. He was well-off enough to indulge his passion in the frenetic late-forties and early-fifties road-rallying scene, where heroic crews would tackle unbelievably-long events like the Liège-Rome-Liège or the Mille Miles – rallies with thousands of kilometres of stages, on open roads, crews wearing tweed jackets and neckties, and, if legend is to be believed, occasionally popping the then-widespread amphetamines left over in France after WW2 by Wehrmacht forces to stay awake. Jean Rédélé proved himself to be a very capable driver, winning several events at the wheel of his decidedly un-sporty Renault 4CV. But he was also a very astute businessman – intelligently leveraging his family background, he had persuaded Renault to grant him a full dealer's concession at a very young age. He was a tall, good-looking and apparently very charismatic man – picture a burlier, French James Dean, cigarette clamped in his lips as he stares at the camera with a Gallic smile and a leather bomber jacket.

Rédélé quickly outgrew his little 4CV – despite hot-rodding it by adding his own design of five-speed close-ratio gearbox, he wanted something faster. Instead of buying someone else's sports car, he bravely decided to establish his own company. In 1955 he built the first production Alpine – the A106, an elegant little coupé with 4CV mechanicals, but with a much better performance than its donor car, thanks to its feather weight (less than 550kg or 1213lb) and excellent aerodynamics, both the result of the lightweight glass-fibre bodywork (high-tech for the mid-50s) draped as closely as possible over the mechanicals, leaving just enough space for the driver and passenger – as long as they were no taller than about 6ft (1.8m). The cars were built in Rédélé's hometown of Dieppe, on the northern coast of Normandy, a pretty, sleepy little fishing port, famous up to then only for the ill-fated commando raid by the Canadian forces in 1942 – a bloody and tragic preparation for the more successful D-Day landings further to the west two years later. In 1962 the Dieppe plant began to build the A110 – the car that would create the Alpine myth. This achingly-beautiful little car – and it really is little, only 3850mm (12ft 8in) long, 1520mm (5ft) wide and 1130mm (3ft 8in) high, hardly troubling the scales at 790kg (1742lb) for the very heaviest versions – early stripped out ones could be as light as 565kg (1246lb). The car was and is known as the *Berlinette* and is so famous in France that saying *Alpine Berlinette* is superfluous – the 'Alpine' is implied.

It's hard to explain how much impact this car was to have on the French psyche. Less than 7600 were built – a tiny number – but it seemed to sink deep into the affections of anyone French. A few factors may account for this – its use by the French Gendarmerie as a high-speed pursuit car, its incredible *palmares* in all forms of competition, but especially in rallying (the iconic *Berlinette* image is of this tiny car, spectacularly sideways on a snowy stage of the Monte Carlo in the mid-1970s), its beautiful, organic shape – but I think in large part it was due to the French love of the underdog. The story of this plucky little family-owned business from Dieppe, thumbing its nose at the larger car companies, not only in France but also its richer and more famous German and Italian rivals, simply appealed to the slightly contrarian French psyche. It would remain in production until 1977, and win a list of motorsport titles far too long to list here.

But by 1973, Alpine was in difficulty. The oil crisis hit sports car sales hard – ironically for a company that was producing highly fuel-efficient vehicles with small-displacement, relatively parsimonious engines. Rédélé was forced to cede majority ownership – and hence control – of his company to Renault, his major stakeholder and source of many key components, including engines. Even to the present day, this deal is highly contentious in France. There is a hard core of Alpine fans that will passionately explain to you – at length and with much gesticulation – that this was a Faustian pact, and that *Le Régie* (an affectionate but double-edged nickname for Renault) deliberately killed the company. Others will explain – equally passionately, but probably slightly more calmly, that Renault in fact saved the company and allowed it to operate for the next 20 years. I will discreetly slip sideways out of the discussion to reserve my own opinion and allow it to be endlessly debated in bars across France.

Whatever the cause(s), the company stumbled on until 1995. The iconic A110 was replaced by the more angular A310 – actually Alpine's most successful vehicle in terms of sales, then by the GTA, and finally the A610. By the mid-90s the company was effectively bankrupt, its resources spread too thin by a wildly eclectic motorsport campaign that saw Alpines competing in almost every imaginable form of motorsport, by an ambition to rival its richer and deeper-pocketed competitors, notably Porsche. Its image was blurred by a nagging confusion between the Renault and Alpine brands – the company could never decide if Alpine was a separate brand, or merely a rebodied Renault, and this brand confusion is obvious in the products and company communication right through the '70s and '80s and into its slow decline and eventual death. But finally, Renault pulled the plug in 1995, the brand was no more, and the Dieppe plant was converted to build the more sporting cars in the Renault stable – hot-hatch Clios, Méganes, and (notably) the little Renault Spider in the late '90s. The plant slowly declined as the investment tap was gradually shut off ... but Alpine lingered in the memory of its many fans.

Renault could not forget Alpine, either. It lingered in the minds of some of the employees: glorious memories of champagne-soaked victories at Le Mans, of vivid blue little coupés ripping open the silence of the winter Alps with ear-splitting straight exhausts, blue exhaust flames lighting up the snowy darkness. It was a romantic dream, a fantasy to cherish in the far less romantic day-to-day task of building cheap and reliable family Renaults.

I am told that between 1995 and 2013, there were up to nine separate attempts to resurrect Alpine. Some of them were mere designer's sketches, getting no further than felt-tip drawings. Some went as far as building physical mule cars. But all failed – the cold, hard economics of building low-volume sports cars would not pass the stringent economic tests of a serious volume carmaker with shareholders to satisfy, including a French government that would brook no waste of tax-payers' money on gearhead fantasies. But Carlos Tavares was one of those who would not, or possibly could not, let go of the dream. Born in 1958, he was of the generation that remembered the glory days of Alpine. A genuine car enthusiast and motorsport fan, in 2013 he put in the effort to sponsor a final attempt ... and managed to persuade Carlos Ghosn to actually go for it it.

How? First, by promising to deploy the same cost-control rigour that he had already demonstrated on multiple mass-production projects. The idea was that Alpine would be a separate organisation – leaner and cheaper than the *grande artillerie* (heavy artillery) of Renault's normal organisation. It was to be a classic 'skunkworks[36]' effort – separate a dedicated team from the mainstream organisation, give them a very limited budget, a strict deadline, and most importantly, freedom. Freedom to re-invent the rules, freedom to break them when required. Tavares was persuaded that this approach would allow Renault to produce a genuinely innovative car – not simply a breathed-upon Renault – and would control the costs to an 'affordable' level. And he had the ear – and the trust – of Mr Ghosn.

---

[36] *'Skunk Works' ™ is the official nickname of Lockheed Martin's Advanced Development Programs (ADP) lab, which has dreamed up such revolutionary aircraft as the U2, SR-71 Blackbird and the F-35 Lightning II. It has come to mean shorthand for a small, flexible team split off from a larger organisation in a quest for new, breakthrough ideas ... often on a shoestring.*

But Carlos Ghosn was not about to place a blind bet, no matter how much he trusted Carlos Tavares' judgement. He insisted that Renault find a partner – a company willing to enter a 50:50 joint venture. This was largely an insurance policy – if the enterprise went south, well, at least Renault would only have to pick up half the tab. It was still likely to be a big bill, but half of a big bill is better than all of it. And the company got lucky – or so it thought. Renault was quickly able to find a partner in Caterham Cars. At first this looked like a match made in heaven. Caterham had recently been bought by Tony Fernandes, an extremely-rich Malaysian businessman who had – apparently – made a fortune in the airline business, and now wanted to invest it in motorsport and high-performance road cars. He had bought Caterham and – importantly – managed to head-hunt most of the engineering team from Lotus cars – the crack team who were largely responsible for that modern icon, the Lotus Elise/Exige, with its amazing simple but effective folded-riveted-and-bonded aluminium platform. Very quickly, as befits a skunkworks operation, Alpine-Caterham Cars was established, and the joint development project was kicked off. The CEO of this new entity, 50:50 owned by Renault and Caterham, was a gentleman by the name of Bernard Ollivier.

Bernard was already a legend within Renault. He had been with the company for many years and was a 'Renault guy' through and through. He had been one of the driving forces behind setting up Renault Sport, the sports car division that had to some extent taken up the performance-car torch when Alpine had faded out. As such, he was the 'father' of some of the craziest road cars that Renault had ever built, including the notorious 2001 Renault Clio V6, a truly bonkers little hot-hatch with a massive V6 engine replacing the rear seats – an evil-handling little ball of French creative weirdness. Now towards the end of his career, the CEO position at Alpine was to be what the French call '*le baton de maréchal*' or field-marshal's baton – a final, prestigious position to cap a long and illustrious career.

I had followed all of this with great interest. I vaguely knew Mr Ollivier, having met him a few times during my time at Renault. I now made it my business to meet him again, and straightforwardly volunteered for service. I explained that the ZOE project was finished, and that I was on the lookout for a new job. Bernard, a gentleman to his core, listened to me politely, welcomed my enthusiasm, but made it clear that I was not exactly the first to 'volunteer' for his new team. I could well believe him – he probably had half the staff of the Technocentre sending him their CVs. The Alpine resurrection had stirred the passions of any Renault engineer worth his/her salary. But he assured me that he would add my name to the short-list for Chief Vehicle Engineer candidates, kindly warned me not to get my hopes up, and said that he would mention my candidature to the key decision-maker – Mr Tavares himself. That was more than fair, so we parted on good terms.

By sheer co-incidence, I had the chance to meet Carlos Tavares myself just a few weeks after this discussion. I was, as I've mentioned, on the look-out for a new post-ZOE job, and my boss Marc Soulas and the HR folks explained to me that I was apparently a lucky boy: my search was over – a job had been found for me. A very important job. I was to be named the next Executive Assistant to Mr Tavares. Now, this role in Renault is considered a very prestigious one. The idea is that experienced managers with 'specialist' backgrounds like myself would spend a year or two as the COO or CEO's right-hand-person – their gopher, guard dog and general factotum. In this way, one could build up the all-important *réseau*, rub sober suit jacket shoulders with the big dogs of the C-suite, and generally be groomed for corporate greatness. It was, genuinely, an honour, and a reasonably sure-fire ticket to very senior roles – assuming the person did not screw up, of course. Only one issue – I hated the idea. Sheer arrogance on my part, of course, but I simply could not see myself as anyone's 'assistant,' whether sweetened by that word 'Executive' or not. I was not particularly interested in corporate politics, nor impressed enough by titles and protocol to find them interesting. But … refusing the position would probably be corporate suicide.

Hence I found myself at 7.30am, drinking a short, stiff wake-me-up coffee in the little waiting annexe on the 7th floor of Renault's company HQ in the western Paris suburb of Boulogne-Billancourt. I was early for my 8.00am meeting with Mr Tavares – like 'the other Carlos,' he was famous for being punctual, and Paris traffic is as unpredictable as the weather, so I had left plenty of time. I rehearsed my lines in my head. I had decided that – for once – I would be a good corporate soldier. Mr Tavares would offer me the job. I would swallow my pride and thank him graciously. I would put in my time – a year or two at most – toe the line, be the best damned Exec Assistant he ever had, then move on to more interesting things. That was the logical decision, and I had made my mind up.

At 7.50am, the door to Mr Tavares' office swung open and the man himself stepped out to wave me in. Grey suit, narrow lapels draped over his spare racing-driver's frame, neat tie, shrewd grey eyes, absolutely no BS. I looked around his office – famously bare, nothing at all on the desk except a computer, a single page of blank A4 paper and his pen, neatly aligned with the edge of the paper. I sat. Straight to business, no small talk, the man himself as efficient as the décor.

"So, David, I know that Eric [the current Exec Assistant] has already briefed you on the job. All clear for you?"

I nodded a yes.

"Good. No need to go through the basics, then. You only need to understand one thing." Pause.

"You need to understand that this is a shit job. The worst. 24 hours a day, nothing but 7th floor politics, no glory, not much fun. But you'll learn a lot and probably go on to something big."

I gulped, inwardly. He was not selling this.

"So, David, what I need to understand from you is – do you have a passion for this? Do you really, *really* want this job?"

I looked at him. His grey eyes stared straightforwardly back at me. He was testing me, of course, but he was also actually asking me an honest question. I decided on the spot that I would not do him the disservice of lying.

"Well actually, the answer is no. I don't, really."

"*Ah oui?*" he sat back, and I could see now that he had been playing with me. It had been the right answer. "Why not?"

Abandoning my carefully prepared rationale, I told him the truth – that I just wanted to design cars, that spending a year or two as his assistant would probably bore and frustrate me. I wrapped it in cotton wool a little, of course – the guy was after all the COO of the company, and it was indeed an honour to have even been considered – but I was honest with him. I concluded by asking him what he would have done in my position when he was in a similar role to me, maybe 10 years before – would *he* have taken the job? He was smiling now … and he admitted that no, probably not, for pretty much the same reasons.

"Well, okay, that's settled, then". He looked at his watch. It was not even 8am. He leaned back in his chair. I'd noticed as I came in that the wall behind his desk was filled with a large print of a *Berlinette* – one of the iconic images of the 1973 World Rally Championship winning A110 'yumping' on a stage – the dirt-splattered little blue coupé flying through the air, rear wheels drooping, crabbed inwards as the swing axles cambered under it. Tavares crooked a hand over his shoulder to point at the car, without even turning in his chair.

"I suppose you want to talk about *that*, then," – meaning Alpine, of course.

I nodded, eagerly. For the next 20 minutes or so we talked about the new project, the resurrection of Alpine. He told me that Bernard Ollivier, as good as his word, had indeed mentioned me, had generously sung my praises, but he too warned me not to get my hopes up too high – there were other candidates, and the choice would not be his alone – we had to keep 'the English' (meaning the Caterham half of the new joint venture) happy. Again, more than fair, so I shook his hand and left – to report to my dismayed boss and horrified HR Department that I had 'refused' the job as his Executive Assistant. They thought I was crazy, and that I had indeed committed career suicide. They were probably right.

A short time later I heard the bad news – someone else had got the Alpine job. Jean-Pascal Dauce, an excellent Renault Sport engineer with long motorsport experience and the amazing Renault Mégane R26.2R ultra-hot-hatch to his name had been chosen. I was bitterly disappointed, but could not complain – it was not surprising that Caterham would prefer a proven motorsport 'ace' over me, someone with no background in sports cars or motorsport. Both Bernard and Mr Tavares had warned me not to expect too much, so I bit back my disappointment, put it behind me, and moved on.

And, in theory, moved up. Renault was going through one of the periodic re-organisations that large companies are addicted to, re-jigging the Engineering reporting structure. I was appointed Director of the somewhat awkwardly-named 'Body Equipment and Mechanisms' team. This was a promotion – it was a Vice-President level position, with several hundred engineers reporting to me all over the world, the majority in France, but others in such far-flung locations as Romania, South America and South Korea. The team was responsible for designing a vast and varied range of components – from door seals to window mechanisms, locks and latches, from windscreens and door glasses to wiper systems, airbags, seatbelts, etc ... a whole list of components fitted on all the vehicles produced by the Renault Group – Renaults, Dacias, Samsungs (in South Korea), now even Ladas, following the Alliance's acquisition of that iconic Russian brand in 2007. It was, again, a '*Métiers*' or 'line engineering' function, like my old Body Engineering job back in Nissan, before Qashqai, but now on a much larger scale. It was seen by Renault as a suitable change in career after my time on the ZOE project – a nice balance of people-management to make sure that I did not get too corrupted by the ice-pick focus of single-vehicle projects. My team would provide the manpower for the various projects, being the 'horizontal' strata of the classic two-dimension matrix structure beloved of car companies. The 'vertical' pillar was projects like the next Clio, Mégane or ZOE, which would draw on our expertise and manpower. A typical mid- to high-level Engineering management position in a big car company.

I did my best. Honestly. For a year, I led weekly meetings, monthly meetings, quarterly all-hands meetings. I did my best to explain new corporate strategies. I listened to the woes, career ambitions, disputes, the real and imagined slights of hundreds of people, and played broker between the theoretical policies of the HR team and the realities on the ground. I shook hands – very carefully – with the entire team every day; to forget to shake hands with one of your staff members is a serious slight, in this most traditional of French companies. I made quarterly, yearly and three-yearly budgets, launched cost-reduction initiatives, gave pep-talks and massaged egos. I travelled all over Europe and Asia to meet and try to get to know as many as possible of the folks who worked for me. I learned the subtle arts of dealing with the no less than four official trades unions that represented my teams in France alone – not forgetting the unofficial ones. Strangely, this was one part of the job I actually enjoyed. I set technical strategies and roadmaps for various component families for the next 10 years, met with suppliers, patting some on the back, playing hardball with others. I promoted folks, demoted folks, slid folks discreetly sideways. I led endless quality improvement Task Forces on small issues like wind noise around a front quarter panel rubber seal. I participated in endless career committees. I kept my nose clean and my profile low. But I quickly realised quickly that if you have the professional and mental well-being of 700-odd human beings to look after, you have precious little time to engineer anything, except possibly a few minutes' break for a coffee. I knew it before, but I realised that I was definitively more of a 'projects guy' than a '*Métiers*' guy. Like it or not, I was a better manager of technical projects than I was of human beings.

Meanwhile, I enviously watched the Alpine story unfold, from the sidelines. Unusually in a company where everyone seemed to know everything, not much information leaked out, but the joint venture seemed to be forging ahead at full steam, the team working intensively with their counterparts in Caterham, crossing and re-crossing the English Channel, driving ahead with the most interesting project that the company had tackled for many years. The Alpine team were an elite, chosen few, housed in a separate building in the far reaches of the Technocentre, in true skunkworks tradition. I was deeply jealous, but the die was cast, and there was no point in looking back.

Nine months into my new job, I had to admit to myself that I was bored to tears. I even regretted not accepting the job of being Mr Tavares' secretary ... although if I am honest, that regret was diluted by some relief when, to everybody's great surprise, he left Renault in August 2013. After a remarkably candid interview with the Bloomberg news outlet, in which he declared his wish to one day be 'Number One' (ie, CEO) of a major car-maker, he left the company, to resurface a few months later at PSA[37], Renault's French arch-rivals, headquartered just a few kilometres away from the Renault Technocentre. His loss, coming so soon after that of Patrick Pélata, was another huge blow to Renault – and even more so, to Alpine. Amid these tumultuous changes at the top, I also realised that I missed the structure and discipline of Nissan. Time passed far too slowly as I commuted dutifully back and forth to my comfortable Vice-Presidential desk and my uncomfortably comfortable line-manager job.

---

[37] *PSA, or more correctly, PSA Peugeot-Citroën was the company that emerged in 1976 when Peugeot purchased its ailing rival Citroën, to create a group that was Renault's fiercest local competitor. It would become part of the giant Stellantis consortium when it merged with Fiat Chrysler Automobiles in 2020, still under Carlos Tavares' leadership.*

CHAPTER 13

"Where do I sign?"

June, 2014. I was plodding through some worthy-but-dull work – a good bet that it was an organisation chart, a budget spreadsheet, or some HR documentation – thankfully, I don't remember. My phone rang. *Monsieur* Bernard Ollivier. Polite preliminaries over, he came out with the following bombshell:

"David, don't get too excited yet, but there just might be a job opening for you here at Alpine after all. Would you still be interested?"

I immediately got too excited. My heart leapt. Suddenly, I realised that I had *not* been able to process the disappointment of not getting the Chief Vehicle Engineer job at Alpine a year before. I had wanted it – badly. I still wanted it – badly. Once again, best be honest.

"The answer is 'yes.' And I don't care what the job is. I'm in. I'm still a pretty good electrical engineer if you need someone to do the wiring. Where do I sign?"

He chuckled. "Thought you might say that. Come over and see me when you can, I'll fill you in. And no, it's not just the wiring I need you for!"

Over the next few days, he briefed me – in great secrecy. It seemed that the project was not, after all, going so well with the 'British friends' as Bernard called the Caterham team, with characteristic diplomacy. The technical work was going ahead perfectly, he assured me (I would later find out that this was not quite the case, as we will see), but the business relations were strained, to say the least. At first, this manifested itself in late payment. The business deal was simple, as befits a 50:50 joint venture. Each partner would pay half of the various costs – for manpower, tooling and prototypes, as well as for more mundane things like office space, CAD equipment and software licenses. Caterham started to show signs of difficulty in paying vendors – either paying late, or asking Renault to temporarily pick up the tab – a bit like that friend who always 'forgets' their wallet at the restaurant. After a few months of this behaviour, it was time to call in the auditors, who identified what had somehow been missed during the due diligence checks at the time of signing the partnership – Caterham had no actual cash. Its wealth was largely that of its owner, Tony Fernandes – and his money was mainly paper, tied up in equity of his airline. It simply did not have the cash to pay for designing a car – not even to pay for half of a very frugal sport car.

Things quickly and inevitably became acrimonious. Not helped by the fact that Carlos Tavares, the man behind the project and the deal with Caterham, was no longer there to smooth things out. Relations between the teams went downhill, fast. Fingers were pointed and tempers were lost. One of the 'solutions' was what was called a 'go slow' – the idea of moving forward on the project, but deliberately at a slower pace, to put pressure on the joint management to find a solution. To my eyes, a somewhat curious strategy; I knew of only one way to execute a car project – full throttle, top gear, using speed and momentum to crash through barriers.

In short, the project was in a critical stage and the short-lived Anglo-French marriage was soon to lead to a divorce. The joint venture would be dissolved. A legal separation would be put in place between the two companies. Renault would buy back most of the Intellectual Property (IP) rights, and Caterham would walk away. But Bernard, the wily old corporate fox, had somehow managed to convince the Renault top management that he could have a period of 'some months' (a beautifully vague deadline) to figure out if Alpine alone could make a solid business case – without a partner – and hence convince the Renault top management that we should indeed forge ahead. It had also been decided that a change in management was required – that the team needed to be reshuffled. This was necessary in any case, because Caterham had literally been responsible for half the car – the 'Alpine' part of the team was now only half a team. It needed to be rebuilt, and the powers-that-be had decided that a new technical lead was required. Jean-Pascal Dauce would return to Renault Sport to pick up the leadership of its motorsport programs again. I was effectively being offered his job.

This was delicate. I did not want to offend anybody, but I wanted things to be clear. It was obvious that the project was – thanks to the financial misjudgement with Caterham – in disarray. It would be a tall order indeed to get it back on the rails *and* to hammer out a business model that would be acceptable to Carlos Ghosn. I told Bernard that I would take the job, but on one condition: that I would have full technical and product decision-making authority. This was a very big ask – I was requesting the powers not only of the Chief Vehicle Engineer, but also those of the Program Director – in effect, asking to be judge and jury. Highly unusual in Renault, but I felt very strongly that the only way forward was to have one decision-maker – shared decision-making or (even worse) no decision making would kill the already-ailing project. Bernard and I shook hands, and I joined Alpine Automobiles SAA with the beautifully simple (and wholly concocted) title of 'Project Director.' It was to be the most fulfilling period of my career.

I tidied up my affairs and handed over to my replacement in my *Métiers* job, and left my comfortable office gratefully, carting my few things across to the non-descript building on the far edge of the Technocentre that housed the tiny Alpine team, sandwiched in between two backroom support teams, well away from the prying eyes of the main Engineering campus. I met my new colleagues, the fellow Alpine Directors that were charged with putting the project back together under Bernard's leadership. Antony Villain was the Chief Designer, young, even cooler than most designers, with a warm smile and a great sense of humour. Eric Reymann was the Product guru, intense, passionate about sports cars, and whip-sharp when it came to the customer's needs and expectations. Yildiz Caprak led the Supply Chain team, charged with purchasing the parts and services required to build this unique car: an outsider like me, born and raised in Turkey, but completely integrated in the Renault purchasing world. A little later, Arnaud Delebecque would join us to look after Sales and Marketing – bringing his big ear-to-ear grin, dancing blue eyes and the crazy streak that is required to succeed in a skunkworks or a start-up. And that was it in terms of 'top management.' The four or five of us had the task of putting the Alpine show back on the road.

I quickly got to know my miniscule Engineering team, the orphan kids of the Caterham divorce. They were a little bruised and more than a little suspicious – the past year had obviously not been easy for them. But I quickly realised that I had some gems – albeit some rougher than others. I also detected a problem – we had two cultures within the team. Some of the team came from Renault Sport, used to working on sporty derivatives of existing cars, with a typically 'motorsport' culture – speed sometimes being valued above thoroughness. Some of them came – like me – from *la Grande Maison* of Renault – used to designing cars from scratch, but with a tendency to be slow, ponderous and a little overly-procedural. I immediately detected the seeds of a problem – we would need to establish an 'Alpine' culture somehow – not a Renault Sport, or a Renault *Grande Maison* culture, but a blend of both, with the strengths of both, and, hopefully, leaving the weaknesses behind. Not easy ... but being the eternal outsider would help me, again. I was not a Renault Sport guy, nor was I seen as a mainstream Renault guy – being non-French and a Nissan 'import' originally, I did not really fit that mould. I could hopefully use this 'outsider' image to my advantage here.

With hindsight, I see now that Alpine was in fact, a classic start-up as well as a skunkworks. Of course, we were not a real start-up in that we were a wholly owned subsidiary of Renault – big brother was always there to pay our wages at the end of the month. So, we did not have the same existential fear that makes engineers in true start-ups so hungry and creative. But we did have a genuine threat – if we failed to get the new A110 back on track, and specifically if we failed to present a viable business plan to Carlos Ghosn and the Renault Executive Committee, we were dead in the water. The team would be disbanded, and we would all have to go back to our former, 'normal' jobs. And no-one wanted to do that. We also had the classic start-up problem

– and opportunity – of not having any rules. We could, in theory, do what we wanted – with all the benefits and risks that such freedom brings. Finally, as mentioned above, we had the typical start-up issue of being a cultural melting pot – different technical and business cultures forced together to create both a product and a company in parallel: each of which tends to be convinced that 'their way' is the best, or even the *only* way to do things. We would have to successfully navigate all these issues over the next few years if this car was indeed to emerge from the Dieppe plant some day.

I immersed myself in the technical details of the engineering that had been carried out so far under Jean-Pascal Dauce's leadership. There was some very good news. The fundamentals were excellent. The underpinnings of the car – the platform engineering – had been mainly led by the Caterham (largely ex-Lotus) team, and I was delighted to see that the basic floor structure was very Lotus-esque indeed – using the classic folded, riveted and bonded construction that had made the 1996 Lotus Elise Series 1 one of the most influential car designs of the postwar period. This construction method had several advantages – it was light and stiff for its mass, somewhere between a cheap, conventional stamped-and-welded steel construction and the very expensive carbon-fibre tub that would be the money-no-object choice. It was also low investment. The tools required were much cheaper than the tools required to make a stamped steel structure, or those required to autoclave a carbon-fibre monocoque. Hence the process was perfectly adapted for a low-volume, impecunious sports car maker like Lotus and now, Alpine.

Unlike Lotus, however, the Upper Body or 'top-hat' of the new Alpine would also be made of aluminium. Rather than the SMC[38] or glass-fibre construction traditionally used by Lotus, the entire body structure of our new car would be aluminium. This would give much better geometrical stability and hence better quality than the traditionally rather hit-and-miss Lotus method: we should be able to achieve fit-and-finish, door gaps and paint quality closer to German sports cars than the rather more rough-and-ready British underdog's products. I thought this was an excellent decision, and never even thought of reversing it, even though it was to cause enormous headaches later on. Bernard and I pushed very hard for his aluminium structure to be built in Dieppe, the traditional home of Alpine, rather than 'internally outsourced' to one of the larger Renault plants in the north of France. This was only one of a succession of controversial decisions that we were to make.

The powertrain choice had been a very contentious one, with a lot of infighting between the Caterham and Alpine teams, as well as complex internal manoeuvrings within the Renault-Nissan Alliance. But finally, the decision had been made, after months of wrangling – a new 1.8L single-

---

[38] *Sheet Moulding Compound or Sheet Moulding Composite. Ready-to-mould composite sheets, pre-impregnated with thermosetting resin.*

turbo engine, which would put out a modest 252PS (248bhp) – enough, if the car could be kept lightweight. This was to be mated to a double-clutch gearbox sourced from the famous German transmission supplier, Getrag. Another contentious choice, and one that many of the more hardcore motoring journalists would later bemoan – the lack of a manual gearbox seen by some as a compromise for a focused 'driver's car.' It was a choice that I would patiently defend later in a hundred discussions with journalists and enthusiasts – explaining that a modern double-clutch gearbox is actually lighter than a manual equivalent, when looked at as an entire system of gearbox, clutch, controls, and even the mounting brackets for the clutch pedal and gearlever. Equally importantly, the double-clutch gearbox with paddle-shift gear controls behind the steering wheel was well adapted to 'less expert' drivers. This was always a difficult point to explain – most drivers, certainly most enthusiast drivers, and even some professional automotive journalists, tend to over-estimate their capabilities behind the wheel. I, on the other hand, had been lucky enough to sit beside enough professional test drivers by now to understand that in fact most 'normal' people (like me) have very limited skills. We simply do not have the motor reflexes, vision and mental capacity of professional race or test drivers. When driving cars fast, our mental capacities tend to get overwhelmed. Simplifying the driving task actually allows the driver to concentrate better, go faster, be safer … and actually enjoy the experience. Reducing the gear-shifting task to a simple pull of a lever, as opposed to a relatively complex series of actions required to change gear properly in a manual-gearbox car at high speeds, allows the driver to relax, concentrate, and fully enjoy the steering, acceleration and braking tasks. Explaining this, without bruising anyone's egos or trying to explain to them that they were possibly not the driving demi-gods that they thought they were, was a difficult task. My approach was to personalise it – to talk about my own (very real) limitations as a driver, and to openly admit that I was both faster, and more relaxed, when driving the car near its limits, with the paddle-shift. Thankfully, most folks would accept (and possibly, secretly agree with) these explanations, although a minority of people would continue to heavily criticise the choice. You can please some of the people half of the time, etc …

So, on the one hand, I had inherited many positive things to work with. On the other hand, lots of fires were also burning. The divorce with Caterham meant that half the team had disappeared. The technical teams responsible for the chassis and body structures were gone, overnight. This meant recruiting people – quickly – with the right mindset to pick up the task and run with it. Fortunately, this proved fairly easy – I was inundated with e-mails and phone calls from people wanting to work on the project. I could no longer walk across the central lobby area of the Technocentre without being waylaid by folks wondering how they could get into the team. I was thus able to hand-pick a few of the people required to rebuild and reinforce the team. I also had the less enjoyable task of diplomatically asking some people to leave the team. Some departed amicably, some less so. But over a few months, I was able to form a solid team around me, as I had been fortunate enough to do with previous projects. My right-hand man was Michel Fumex – a Renault veteran, with a solid background in vehicle packaging. The word that comes to mind in describing Michel is 'solid' – he was tall, physically imposing, but always calm and reassuring. Thierry Annequin had joined the team just before me – a brilliant chassis engineer who, like me, had found himself in a less-than-interesting management job, and now had the chance to apply his enormous theoretical and practical skills to designing the chassis of this new car. The car was destined to win countless 'best handling' tests and awards ... these were largely due to Thierry's genius. Hubert Ernault was another rock on which I could build the new team. He had almost decided to become a professional soldier in his youth, and he was the kind of guy with whom you would choose to be in a foxhole. He was a body engineer, with wide experience in both design and production engineering in his chosen discipline. Like Michel, he was calm, unflappable. Fortunately, he was also armed with a great sense of humour – a great asset, as he would have one of the toughest jobs on the project – making this beautiful all-aluminium structure a reality, while keeping to the feather-weight principles of Jean Rédélé and the Alpine brand.

It did not take much management intuition to realise that decision-making was a problem. Or rather, lack of decision-making was a problem. The team had got into a habit of endlessly discussing each technical, commercial and product decision that needed to be taken. These discussions often became emotional and conflictual, and continued seemingly endlessly. There were a number of reasons for this. One was that it was a team of experts, with a genuine passion for their subject. Put five experts in a room and ask them to come to a consensus – on anything. Good luck with that. Secondly, the team was terrified of the pressure. They felt the weight of the brand history, heavy on their shoulders. They felt the pressure of the parent Renault Group. They knew that we were fighting for survival. They were thus more afraid of making a bad decision than of not making a decision, and simply continuing on with increasingly entrenched circular discussions. Finally, a subject that was endlessly discussed, but was in fact the *least* important factor – that there were some structural errors in the management organisation, which made it unclear who had the actual decision-making power – and hence made it easy for people to wait for someone else to make a decision. I immediately saw that I could solve this. I had by now many years' experience of being the person who 'made the call,' and felt very comfortable with the science and art of technical decision-making.

Within days, I made it clear – by actions, not words – that we *were* going to take decisions. Quickly, without fear, and without regret, once those decisions were taken. I also made it clear that the key decision-maker was me – if I made the right decisions, I would share that responsibility (and credit) with the team. If I made a bad call, that would be on me. I could almost see the physical relief on the faces of the teams – here was someone willing to make decisions, and (more importantly) to live with the consequences. I had by now studied the theory and practise of decision-making in detail, and I never felt the least uncomfortable with this role. Starting to make decisions, *keeping* decisions once made, and then acting on them, was like putting fuel in the tank of a sports car: things immediately started to move. Quickly.

In parallel, I tried hard to create a new engineering culture within the team. I made it crystal clear that I did not care about internal politics between Renault Sport, Renault, or anything else – we were all one team now, and what we had done yesterday, or what we might do tomorrow, was irrelevant. I clamped down hard on any internal silos, and very hard on any form of intellectual elitism or engineering snobbery. I made it clear that this was not a loss-making vanity project where we could just play at making sports cars. We were going to build a business that stood a chance of making money, and purism would take second place to pragmatism. Of course, this was not universally popular. Many of the more 'purist' engineers probably considered me a mere bean-counter, with a background in low-cost boring family cars, with no pedigree at all in 'real' sports cars or motorsport. This was in fact true – and several people left the team as a result, to be replaced with more pragmatic production-oriented engineers, who were motivated by getting products across the line. Egos were bruised, complaints were made, but our chances of actually getting the car into production dramatically improved. I started to realise that Alpine, more than any other project that I had worked on, created strong emotions. I would make some new friends on this project, but I seemed set to make enemies also. It was a price I was prepared to pay, if necessary.

'Slow mode' was over. We metaphorically buried the accelerator pedal in the (lightweight) carpet and drove the project forward. Once again, the first step was building mule cars to prove the basic concept. These were based on cut-and-shut Lotus 'tubs,' but with the suspension geometry of our new vehicle, and the all-new powertrain bolted into place amidships. These cars were built in the competition workshop at Dieppe – the first 'real' Alpines to be built there for many decades. It was an unspectacular but historic moment – the team gathered in February 2015 in one of the deeply-unimpressive low-roofed, seagull-shit-splattered sheds at Dieppe, while the build technicians put the final touches on the tiny, low-slung mule. It looked like a very badly repaired Lotus Exige after a serious crash – with hand-made bits of ripply glass fibre pop-riveted on, and a quick blow-over of matte black paint. But the dozen or so of us present for the event knew its true significance, as the starter motor whirred and the little four-cylinder engine barked into life, its busy industrial chatter reverberating off the walls through its hand-made exhaust. This car was a moment in history ...

In parallel to building the first prototypes, we worked through the task of selecting the suppliers. This was a very interesting process, very unlike the previous mass-production projects I had been responsible for. Qashqai and ZOE were high-volume programs, fully supported by their respective mother companies. The successful suppliers bidding on those projects knew that they would be able to supply hundreds of thousands, if not millions of units of the components they were selected to supply. The contracts were enormous, in financial terms, with the full backing of the Renault-Nissan Alliance Purchasing organisation behind the engineering teams to identify a 'panel' of several competing suppliers, evaluate their offers, deal with the complex legal and commercial terms, and negotiate aggressively to get the best deal possible. Even the smallest contracts for a bolt or a tiny plastic clip would see several suppliers competing actively for the business. We did not have to go out and find suppliers; they came to us – it was a relatively simple task of filtering the offers to select the best technical and commercial offer.

Alpine? Very different indeed. First, we were building a few thousand units a year – not hundreds of thousands. Secondly, we did not really have the full support of the enormous Alliance Purchasing team, who could not reasonably be distracted by this minor project. Finally, suppliers were not stupid – they could see that this project did not (yet, at any rate) have the full support of the company. They knew *something* had happened with Caterham – not the full gory details, but enough rumours had leaked to be worrying. Suppliers are used to car companies flirting with 'halo' sports car projects, before deciding that they don't make economic sense and simply giving up. Suppliers often get burned by these vanity projects – having committed their engineering and commercial staff to making offers, they can end up with no business. It was therefore extremely difficult to find suppliers willing to work with us. Many major suppliers, with huge turnover with the Renault-Nissan Alliance, simply turned us down. Yildiz' team did an amazing job of finding alternatives – going out to the market and finding companies willing to deal with us. Many of these were smaller companies, often with a background in low-volume luxury or sports cars. Quite a few were based in northern Italy, in the Po River basin around Turin, the traditional heartland of the Italian suppliers who supported the low-volume and often cash-strapped Italian exotic manufacturers through their regular cycles of boom and bust – companies used to low volumes and high risks, with a proud tradition of craftsmanship and technical expertise. In many cases, these were the only suppliers willing to take the risk of dealing with us, and willing to accept the very tight budgets with which we had to operate. We were very grateful to find any companies willing to work with us. But this also sowed the seeds for later problems ...

In contrast to the struggle to find suppliers willing to work with us, the work in the Design studio was refreshingly easy. Antony Villain's team was also a small, skunkworks outfit, just a handful of very talented designers working in a small annexe, separated from the main Renault design studios, with a discreet Alpine 'A fleché' or 'arrowed-A' logo proudly displayed on the electronically badge-restricted door. Unusually in the modern car industry. Antony and his team controlled everything visual to do with Alpine – not just the shape of the car, but the logo itself, the 'look' of the website, the merchandise, even the livery of the endurance racing cars that we sponsored at Le Mans. That gave the brand an unusual consistency from day one, and helped to ensure that we did not repeat the errors of the past – confusion with the Renault or Renault Sport brands.

The foundations of the design were as sound as the foundations of the platform – Antony and his team, working closely with the vehicle architects, had shown the discipline to keep the car small – just 4178mm (13ft 8.5in) long, 1798mm (5ft 11in) wide and a hip-scraping 1252mm (4ft 1in) high. Tiny, by modern standards. They had crafted an aerodynamically efficient and very 'pure' shape. We were very proud to be able to engineer the car so that it did not need a rear spoiler to keep it stable at high speed. The clean lines of the rear trunk lid, acting in conjunction with the rear diffuser, would generate enough downforce to keep the rear axle pinned, without the need for any aero addenda. The form of the car was simple, elegant and unfussy. Antony was able to resist the temptation to 'over-design' the car, keeping the volumes and graphics very simple. Hardest of all, he and his team were able to clearly evoke the original, 1970s Berlinette – that icon of French automotive history – without resorting to design pastiche. This is a very delicate balancing act, and many teams had toppled off this particular high wire: attempts to ape the VW Beetle and even (dare I say it) the Mini, had inevitably ended up as cheesy neo-retro replicas, clumsily reproducing chrome and toggle switches and cynically playing with customers emotions – sometimes to great commercial success, let it be said. Antony's design of the 'new' A110 was truly respectful of the original – aiming to reproduce some elements, but subtly and with taste. A dab of the original's perfume behind the ears: not spraying it all over the place.

As the project gradually took shape, mules tearing around the test tracks, suppliers coming on board, and the final shape of the car being fine-tuned on the clay models, it was becoming clear just how controversial this project was. It was regularly discussed in the French automotive media, speculating endlessly about what the vehicle might be like … and generally tending to write the project off before we had produced anything at all. There was huge pressure – both positive and negative – from the various Alpine enthusiasts' clubs, many in France, others worldwide, in countries like the US, Australia and Japan. Internally in Renault, there was a strange schizophrenia about the project. Many people – including some very senior executives – felt that Alpine was a mere vanity project, a waste of rare financial and human resources: that we should be concentrating on the high-volume products, on EVs like ZOE, and that if we wished to have a Marketing or 'sporting' presence, that we should simply invest money in Formula One. All very valid points of view. There was also a strange fatalism or lack of self-belief – a sort of conviction that the project could never succeed. Even if we could design a truly lightweight vehicle (which most people thought we could not), that it could somehow never be as good as a German sports car like a Porsche. It was French, right? The French could not make sports cars – everyone knew that! Ironically, this negativity was expressed most strongly by the French executives themselves. Hence the project was considered by some to be a pariah – a waste of money, doomed to failure. A career killer – anyone with any political sense, or ambition to build a reputation, would stay a million miles away from it.

And yet the very people who were most vocal in criticising the project seemingly could not resist 'playing' with it, or attempting to tell us how it should be done. I constantly butted heads with senior executives – many of whom had frankly no knowledge, appreciation of or interest in sports cars – 'advising' us on details of the project and suggesting various 'improvements.' In most cases I was able to politely ignore their well-meant suggestions. In a few cases I would have to remind them that it was an Alpine, not a Renault, and that we would not be taking their advice. This raised many hackles. As did some practical details. The fact that the Design studio was restricted, even to senior Renault executives, was deeply contentious. Requests to drive the prototypes were regularly refused. This was not simply because we had very few prototypes (far fewer than a normal Renault project would have), or because they had in fact no need to drive them. It was also because it could be potentially dangerous. Relatively powerful, light, short wheelbase, mid-engined prototype cars with no ABS and no ESP[39] are potentially dangerous weapons to people who are not used to them. And many of our senior executives had never driven a powerful rear-wheel-drive car. It was difficult to explain to these (typically) alpha-male senior executives that they could not drive our prototypes: for their own safety. I had suspected that the Alpine team would not make many friends. I was right.

---

[39] *Electronic Stability Program, also known as Electronic Stability Control (ESC). An electronic 'guardian angel' that can cut engine torque and activate the brakes on one or more wheels to try to recover the car when it seems that it is about to spin or otherwise have a 'departure from controlled flight.' Early prototype cars usually do not have such systems, as they are rarely used on public roads, and are driven by trained professional test drivers.*

CHAPTER 14

Pressure game

16th April, 2015. Bernard Ollivier and I were standing in a dimly-lit corridor at Renault Headquarters in Boulogne-Billancourt, waiting to go into the Executive Committee meeting that would decide the fate of Alpine. To say we were nervous would be an understatement. We knew that we had the best chance of resurrecting this French icon since its death in 1995. The level of expectation had been inexorably ramped up by the media and the various Alpine *passionnés* (enthusiasts). And this was not a mere rubber-stamp presentation to get the Exec Committee's blessing. This really could go either way. We looked at each other and giggled nervously, like schoolboys before a tough exam. Then the door opened, the previous presenters filed out, we put on our game faces and filed in. The future of Alpine would be decided in the next 45 minutes.

Quick rewind. When the hasty Caterham marriage ended in divorce, Carlos Ghosn had insisted that Alpine could only continue if it did not lose money. It did not have to be a huge money-spinner, but it could not be a money-losing halo project, justified, like the Formula One team, only on the endlessly debatable marketing benefit to the other Renault-badged products of the company. It had to break-even in its own right. Even though Carlos Tavares – by far our strongest executive sponsor – was gone, we had to show a zero-loss business case to the remaining Carlos: now the only Carlos who counted. Since joining Alpine a mere nine months before (which, in true start-up fashion, already felt like nine years) I had not only been leading the technical teams engineering the A110, I had also been building the case for how to make Alpine profitable as a long-term business. I will not go into the confidential details, but suffice it to say that we had built a plan for several vehicles. It was clear that the little two-seat A110 alone could never support a full business and a dedicated dealer network. It was to be followed by other vehicles – all Alpine badged, all unique designs, all with cutting-edge technology, but (crucially) keeping to the core Alpine value of agility: meaning aggressive mass control would remain at the core of all future Alpines. It was an ambitious plan and required significant investment – several multiples of what the A110 alone would cost. And yes, it was risky – pouring significant investment into an all-new premium sports car line-up, before we'd launched the first car to wear the badge, would require a lot of corporate courage. I had worked for weeks on the numbers, checking, and refining the classic COP (Consolidated Operating Profit) and NPV (Nett Present Value) numbers obsessively, making sure that our assumptions were 'centred' – neither crazily optimistic nor overly pessimistic. The numbers were tight – there was no fat in any calculation, and little margin for error.

As Bernard and I took our places at the Renault board-room table, I had flashbacks to the Nissan Executive Committee all those years ago that had cancelled my very first project – B32A. There he was – the same Chairman: Mr Ghosn himself, still sitting at the head of the table, 13 years later and on another continent. A little plumper, different hair, but those same dark eyes and black inverted-V eyebrows, lowered in concentration. The Japanese have an expression *kuuki wo yumo* or 'reading the air' – meaning the ability to sense the atmosphere in a meeting. It was not difficult to read the air in this particular meeting. It was tense, to say the least. Everybody knew that this was a difficult decision. The Alpine project was a little bit like an unexploded bomb. If we succeeded in defusing it, we would be heroes, and everyone would clamour to be associated with it: "Yes, I remember it well, I was there – I passed them the pliers and advised them to cut the RED wire!" But if we failed ... well, it was best to be well outside the blast radius.

We presented the deck – no longer a sheaf of acetate transparencies like back in Japan in 2002; these days it was the ubiquitous PowerPoint presentation. A very short, cut-to-the-chase presentation as Ghosn liked – the chase being the business case numbers. Everyone could see that it was a margin call. The margins were very slim. We were done with the slides in a few minutes. Ghosn thanked us, as polite as ever. My heart then sank. It was 2002 all over again – he asked for everyone's opinion, around the table, one-by-one. It took a lot longer than the last time I had endured this ritual, more than a decade earlier – the Renault Executive Committee was a lot bigger than Nissan's, and as this was Alpine, everyone had a lot to say. As each executive spoke, I wrote down two columns of names in my little notebook: one was titled 'with us' – the other, 'against'. But most people were careful not to commit too much – staying as neutral as possible, without appearing too lily-livered: just giving the merest hint of their position. A few had the courage to say openly and clearly that we should stop the Alpine project there and then. Even fewer had the courage to strongly support us. As always, it would be Ghosn's call. Bernard slipped a sideways glance at me as the CEO of the Alliance summed up. He was as nervous as I was.

Ghosn summed up in what was, for him, an unusually long closing statement. To summarise, he gave us a very conditional approval. He made it clear that he was – I quote – 'reluctantly' approving the project, and listed out a long list of conditions. Again unusually for Mr Ghosn, he stumbled over the word 'reluctantly,' and had to repeat it three times to get the pronunciation right. Finally, he fixed Bernard and me with that famous gimlet-eyed stare and made it clear that it was *our* responsibility – the executives of the Alpine team – to make this work. The message was clear – there was no safety net. If we screwed up, we were finished. It was therefore, unfortunately, a deeply ambiguous 'Okay'. But it was an okay. Bernard and I stood up, thanked the group, and got the hell out of there, leaving the next decision-seekers to take their places. In Bernard's excellent book about the Alpine project[40] he notes that the time was 3.17pm. Alpine was back – officially.

Back in the corridor, our joy outweighed our misgivings. Bernard thumped me on the back and pumped my hand up and down for a long time. "We did it, David! We're still alive" he beamed. It was a genuine moment of victory, and we were wise enough to savour it. But we knew that Ghosn's conditional approval would cost us dear – his clearly and very 'public' less-than-whole-hearted approval of Alpine would mean that we would never have the full support of his powerful lieutenants. They would continue to treat Alpine as the bomb-disposal team – staying close enough to bask in the glory if we pulled it off, but far enough away to not get caught in the blast, if and when we blew it.

Bernard's 'still alive' motto became an oft-repeated bitter-sweet mantra. Over the next two years, we would come close to the death of the project on so many occasions that we lost count. Every time we managed to avoid the worst, often at the last moment, the 'still alive' motto would be muttered through clenched teeth, the adrenaline of the latest near-death experience coursing through our veins.

Throughout the project, we did our utmost to keep the core team small and concentrated. There was an obvious reason for this – the development budget was incredibly tight. Well less than half of what Qashqai had cost in terms of investment – and that was already a pretty frugal project. With this budget, we had not only to engineer the vehicle, renew the neglected and rather dilapidated Dieppe plant, but also establish a small network of Alpine dealers. There was no way that we could reasonably sell a sports car that would inevitably be compared to products from prestige brands like Porsche from the corner of a showroom selling family Renaults and cheap-and-cheerful Dacias. Engineers are expensive – keeping costs down means keeping the number of engineers and technicians on a project as small as possible. But financial reasons aside, we also wanted to preserve the skunkworks ethos as much as possible – to make the team think of different solutions, ignore the rulebook and be as creative as possible.

Partly for this reason, the engineering team was moved out of the impressive Technocentre and into a dedicated annex in the Renault Sport

---

[40] *'Alpine, la Rénaissance'* by Bernard Ollivier, published by Solar Editions, ISBN 978-2-263-16055-4. *Highly recommended to those who can read French. Tells the tale of the company and its revival with a lot more detail on communication, marketing and motorsport efforts.*

facility in Les Ulis. This was a decidedly less impressive location – Les Ulis is an unprepossessing sprawl of industrial units, shopping malls and DIY stores in the scruffy smudge of suburbs to the south of Paris. The Renault Sport buildings are work-a-day low-rise steel-framed prefabricated offices and workshops, and Alpine occupied a one-storey, almost windowless glorified shed on the site. No name on the door, no sign that we were there – in true skunkworks tradition. It was in this deeply unflashy building that we would try to create something to compete with the giants of Zuffenhausen, Munich and Stuttgart. One great advantage of this location was that virtually the whole engineering team could be housed under one roof – the core Alpine team, as well as the sub-contract engineers that supported us. The Renault Sport engineers who worked in parallel on the 'hot' Méganes and Clios were just a few yards away, making it easy for us to call on their help and support. The Technocentre was only a half-hour drive away, if we needed the support of the heavy artillery. And the historic Dieppe factory in whose corridors you could still almost hear the footsteps of 'Monsieur Jean,' as everyone still referred to Jean Rédélé, was an easy two-hour drive away. So we had a concentrated physical green-housing of the team – very useful to maintain team cohesion.

Successful projects need a clear focus. On Qashqai, the focus was on ruthless cost-engineering. On ZOE, it was on making cutting-edge new technology affordable: bringing new EV systems to the market in a safe and reliable way. On the new Alpine, the focus was crystal-clear: it was on mass – or rather, on the avoidance thereof. Alpine's DNA could be summed up in one word – agility. Alpines had always been relatively low-powered, small, and very, very light. This lightness gave them their characteristic feel – and drove their motorsport success. Alpine has often been portrayed as 'a French Lotus,' and Colin Chapman's over-used "Simplify, then add lightness" quote could indeed have come from the mouth of his contemporary, Jean Rédélé. Our challenge was to control the mass of this new Alpine, while keeping the car useable every day.

It is relatively easy to engineer a very light car, but at the cost of useability – take out the air-con, sound isolation, nice seat materials, audio systems etc. The resulting bare-bones vehicle will doubtless appeal to purists, but it relegates the car to weekend or track use only, and severely restricts its market appeal. For that reason, we did not try to chase the tempting round-figure 'sub one tonne' (1000kg or 2200lb) target – a nice-sounding but completely arbitrary number. Instead, we allowed the car to push slightly over a tonne (it ended up at 1080kg, or 2381lb) but, keeping a reasonable level of creature comforts – functioning air-con, reasonable levels of sound isolation and hence NVH[41], nice materials such as leather and a smattering of

---

[41] *Industry jargon for Noise, Vibration and Harshness. Better known that its other three-letter acronym cousin, BSR, for Buzz, Rattle and Squeak. Engineers spend a lot of time battling these six little letters.*

soft plastics and plush carpet, a (rudimentary, and frankly clunky) navigation system etc. But the target was still hugely ambitious – almost 300kg (660lb) less than the car it would inevitably be compared to, the Porsche Cayman. That's three very large adults, or four average-sized people. Achieving this required the same discipline as achieving the cost targets on Qashqai – but dealing in grams, not euro-cents. All decisions were scrutinised for their mass impact and grams were counted as eagerly as a drug addict counts their daily dose. The Press would later enthuse about the level of detail into which we plunged, geeking out over details like the tiny jewel-like aluminium brackets holding the brake lines, engineered to save 5g (⅕oz) each.

As the team got into the detailed design, system by system, we hit our first major technical issue – and the problem was mass. As I've mentioned, the original platform concept had been sketched out by the Caterham side of the original joint venture. They had laid out a delicate, elegant, ladder-like system of simple aluminium extrusions, beautiful to look at on the CAD screens, light, and theoretically cheap to make. Only one problem – it didn't work. As we rebuilt the body and chassis teams, and the new engineers started to look deeper into the calculations and simulations, it became clear that the structure was simply *too* light. We could just about get it to work in crash – by adding some patches, increasing the wall thicknesses of some of the extrusions, and adding a few structural rivets here and there, it could – just about – scrape through the various frontal, rear, side, and rollover crash tests. But what it could *not* do was reliably pass the durability tests.

When we simulated the accumulated damage from road inputs – the stresses and strains of the suspension members bumping over uneven surfaces for the next 10 years – the results repeatedly predicted that the structures would crack. Various more sophisticated models of how the all-aluminium body would behave in torsion and longitudinal bending gave the same tests – the car would be fine for, maybe 5 to 7 years. Hell, it might even be fine for 10 or 15 years if the owner happened to live in a country with nice smooth roads. But some owners, in some markets, would start to see micro-cracks developing after 5, 6 or 10 years. It was tempting to ignore what the workstation screens were telling me – those little red and black areas lighting up in the FE[42] analyses.

It was tempting to ask Hubert's body team to simply stick in a few more patches, add rivets, double up on the adhesive bonding beads here and there. To be honest, it's probably what a real start-up would have done – cross your fingers, go ahead anyway, and deal with the warranty claims in ten years'

---

[42] *Finite Element analysis – a mathematical modeling technique where complex shapes (like parts of a car's platform) are broken up into a mesh or 'net' of tiny elements, each one small enough and geometrically simple enough to allow its mechanical behaviour to be accurately predicted. Stitching the mass of the results back together gives an accurate prediction of how large, complex items will behave. The workstations that run this software will show highly-stressed areas in colours like red, purple and black. Under-stressed areas will show as comforting greens and light blues.*

time. But I could not, in all professional conscience, do so. It was not right, and we had to fix it – properly. This issue become known as the 'K1' – K1 being the highest severity of an identified technical issue – a 'job-stopper' in the jargon. A K3, on the other hand, would be a minor issue, to be solved when we'd have time to get around to it.

So I did what the Chief Engineer must sometimes do – the equivalent of standing up in the train and pulling the emergency cord: I stopped the project, so the whole team could concentrate on solving this K1. The Start of Production would now slip back, one day lost for every day it took to solve the issue. This is a moment of truth for a Chief Engineer. Everybody – and I mean everybody – will have difficulty with this. Of course, people will say that they want the problem fixed, and that you were 100% right to stop things to fix it properly. But they will not want to accept that it will take more time. Executive after executive took me to one side to say "Great job, David. You did the right thing. We should do things right – we need to fix this the right way". They would then look around conspiratorially to make sure no one was listening "*Mais, dis-moi ...* you'll be able to recover it, right? I mean, the timing? We won't have to delay the launch or anything, right? You and your lads will catch it back, won't you?"

This is where you need a bit of cold blood. I'd look them in the eye and say no – we won't. I'd explain that a day lost is a day lost ... we haven't yet designed the time machine that would be so useful on these occasions. Of course, this was a little disingenuous on my part – I suspected that we would be able to catch back *some* of the time lost – for every five days lost, maybe we could find a way to catch back two, for example. But I was trying to make the company realise that there were no miracles here – if we wanted high quality, something had to give – money, or time. It's the old saying of those that have spent a lot of time managing that tough ol' Iron QCD Triangle – you can pick any two. But you often cannot have all three ...

The body, chassis and packaging engineers did a great job. They threw out most of the original 'Caterham' design – although of course sticking with the basic material – aluminium – and concept – bonding and riveting – and designed a more complex, unfortunately slightly heavier, but now properly robust underfloor structure that preserved the suspension geometry of the original, performed much better in a crash and, crucially, would now last for decades, guaranteeing a great quality product not just in the showroom, but for decades to come. The K1 crisis was past – but it was far from the last.

The ambiguity of the April 2015 Executive Committee conclusion (the 'reluctant' and conditional support of the CEO), together with delays and increased costs as we worked through and corrected some of the, let us say, 'loose' early design work, started to worry some of the Renault senior executives. We were under constant review – can we really make this car? Will it really hit the cost targets? Will it be delivered on time? We spent inordinate amounts of time presenting progress, reassuring people, repeating the agility-means-mass-reduction mantra again and again. But it was obvious that the majority of the Board remained sceptical, and would have preferred to simply stop the Alpine project, especially now that its originator, Carlos Tavares, was running the opposition – PSA – and, somewhat worryingly, seemingly making a great job of it. Something had to be done to get these sceptics back on-side. Luckily, we had a secret weapon – our development mules. These were, as I've said, ugly, matte black, cut-and-shut prototypes, with zero creature comforts – they were noisy, basic working tools. But we decided to put a few key Executives in them – to allow them to drive and, crucially, to be driven in them.

This was very risky. These gentlemen – sorry, still no female executives – were not used to being exposed to early-stage prototypes, let alone to early-stage prototypes of hardcore sports cars. They were also not used to driving mid-engined, relatively high-powered and very light sports cars with no electronic safety nets – difficult, demanding things to drive. But Bernard Ollivier and I had an inkling that the risk was worth taking – that the cars themselves would convince these people whose support we needed, whether we liked that fact or not. We accordingly set up an event at the Aubevoye test track where we invited Thiery Bolloré, now the number two of the Renault group, and another key executive, who then led Renault Group's Sales & Marketing division – to drive one of the mule cars. One by one we strapped them into the mule's seats – at first alongside our test drivers, then in the driver's seat. They were some of the first non-Alpine people to experience the sensation of these amazing cars – the brutal acceleration, the huge grip levels as the test drivers drifted them in clouds of tyre smoke through the handling course; the mind-numbing noise of the virtually-unsilenced engines a mere foot or so behind their (helmeted and ear-plugged) heads and, most important of all, the amazing agility and ride comfort of the vehicles. Of course, we had the competitors available to show the differences back-to-back: the dramatic but disappointing Alfa 4C, the track-scalpel sharp, but deeply impractical Lotus Exige, and the inevitable, endlessly-competent Porsche Cayman – so capable in fact that it flirted with that killer adjective 'boring,' hence opening the door a crack for us. But the little dumpy black Alpine mules already showed their unique ability to flow and dance, especially on broken surfaces.

It worked – at least partially. The execs got out of the cars red-faced, sweaty, with very un-executive tousled hair and big grins. We had two key supporters at least. But as always there was a catch. The Sales & Marketing exec – who happened to drive a Porsche 911 himself – pointed at the little mule, ticking and tinging behind him, stinking of burned rubber and carbonised brake dust. "David, if you can build a series production car like that – just like that – you will have a great success". I smiled. This sounded good. But then the kicker. "But you won't. It will get heavier, slower. The quality won't be there. Renault can't deliver such a car. If you could, I'd support you. But you won't. Sorry".

I gritted my teeth, smiled and gave the usual response "Ah, well. Lucky that it's not a Renault, then. It's an Alpine, right?" This was to be a constant partly tongue-in-cheek reply, all the way through the project. In fact, we lacked confidence – in ourselves. The French car industry suffers from a deep inferiority complex, constantly comparing itself to the German competition. And frankly speaking, this lack of confidence has been justified many times, at least in 'premium' and sports cars. But it seemed to be an almost fatalistic attitude – that because we failed in the past, we were condemned to fail in the future. This, I did not accept – I thought there was a genuine chance with Alpine to rival the German incumbents – not by trying to copy them (that would indeed be suicide) but by offering something subtly different – something lighter, in all senses of the word. But very few people within the Group shared this confidence. It would be a constant struggle to shore up confidence, to endlessly convince, and re-convince the company that the risk was indeed worth taking. At times it was an exhausting fight – far more difficult than engineering the car. But our best weapon was always the car itself – we would repeat this exercise many times, by inviting various more-or-less sceptical senior execs to the test rack, putting them into the cars, and watching their scepticism melt under its charm. A little like a vaccine booster injection, we'd have to repeat these events every six months or so to stop them wavering … and so the delicate dance would continue, all the way through the project. I should mention, in fairness to this particular executive, that he *did* support us all the way through the project, as we proved that we could in fact control the mass of the car and preserve – or even improve on – the attributes of these first demonstration mules.

The senior management changes continued. In February 2015, Bernard Ollivier was replaced as Managing Director of Alpine by Michael van der Sande, the head of Sales and Marketing for the Renault Group. Michael was a tall, laid-back Dutchman, who did not have the typical Renault 'lifer' background, having travelled the world working for companies like Aston-Martin and Harley-Davidson. He had been one of the few staunch supporters of Alpine from the very first day, and was now handed the top job, as the market launch of the brand loomed closer. I was delighted that Bernard would stay on, on paper as Michael's deputy – this would make sure that *Le Patron* would get the opportunity to see the car come into reality before taking his well-deserved retirement.

As well as keeping the senior Renault Group management from flinching, supplier quality was another serious problem. As I've explained above, we could not simply rely on the standard Renault-Nissan Alliance mass-production suppliers – even if they had been ready to do business with us, their business was based on the high-investment, low unit price model of all mass-production cars – the precise opposite of Alpine's. Hence we had been obliged to work with suppliers from the world of low-volume prototyping, or from the low-investment high-value world of brands like Aston-Martin or Ferrari.

Stamped metal components were a particular headache. The rear wing/fender of the A110 was one of the trickiest parts to get right. This beautiful, swooping, one-piece panel was a true work of art, its complex curves designed with love by Antony Villain's design team, the shape of it driven by our desire to eliminate any 'bolt-on' spoilers – a beautiful, industrial yet organic shape. But fiendishly complex to stamp. The press simulation tools showed it couldn't be done – the computer modelling of what would happen when you tried to press such a deep shape showed that the metal crystals would rip apart and the metal would tear right at the 'point' of the fender. Red light – no can do. I reviewed the data with Hubert, my trusty body/press expert. The zone where the fender 'lit up' in red was very small – a few millimetres long. "What do you think, Hubert?" I asked, leaning over his shoulder and peering at the screen. Hubert had been a key guy in solving the K1 issue a few months before, showing him to be cool under the heaviest of fire. He had decades of experience, was bold but not crazy. I trusted him as much as I trusted the sophisticated workstation running state-of the art software in front of us.

"Hmmmm. In theory it's a no-go. But ... I dunno. Might just work. It'll be tricky as hell, though".

In short, we went for it. More skunkworks rule bending. In a normal project, there was no way we would have done this. I would simply have informed Antony that the panel could not be pressed, and that we'd just have added a plastic spoiler – like every other sports car on the market. But this was different – we were there to take risks, push the envelope; to use our experience and judgement, not just blindly apply the rules. This decision would cost us dear – in fact the damned fender *did* rip on stamping – over and over again. It would take up more than a year to get this right, involving multiple tweaks and re-iterations of the press tool. It would be one of the most contentious issues of the project. We would be roundly criticised by our corporate Quality and Manufacturing teams for having taken the risk, and some saw me as a devil-may-care risk taker who wilfully broke the rules. Nothing could be further from the truth. In the end, we managed it, the supplier *did* find a way to press it repeatedly without tearing, and the car saw production with the beautiful, un-be-spoiled rear end. But at the cost of many long nights, endless meetings and, and untold time spent at the beleaguered supplier's site in Italy. Even today, every time I see an Alpine on the road, the rear fender seems to wink at me as that low-slung, simple rear end recedes into the distance ...

As per the Qashqai and ZOE projects, paper drawings and digital models slowly started to become real. Aluminium, steel, plastic, rubber and leather parts accumulated in the Dieppe plant – itself being slowly renovated in parallel – and we started to build the first, badly-named (another hangover from the Caterham days) 'Body Prototype' cars. It was as magical as ever to see these very first Alpines that actually looked like Alpines take form – the sharkskin-grey raw aluminium bodies coming out of the paint shop, resplendent in new coats of Alpine blue paint, then being slowly filled with engines, harnesses, axles, interiors, until we finally had the very first 'real' prototypes (as opposed to mere mules) sitting there. The very first car was stubborn – it took us nearly 48 hours to get its engine to fire – but finally it did, and the walls of the Dieppe factory vibrated to the exhaust pulses of the first real Alpines for almost three decades. Another very proud moment for us all.

Yet more risk-taking. We decided to present some of these very first cars at the Geneva Show, to be held in March 2017, just a few weeks away. This was – again – a total no-no for a normal project. Show cars may not always be full production cars, but they are typically based on very late prototypes or early pre-series cars, with minimal 'fettling' required to make them presentable. We broke most rules in all the books by taking the first bodies and fairly extensively adjusting them to make them as good as the production cars would be – in terms of fit and finish, cosmetic appearance, and the quality of the characteristic 'Alpine blue' paint – a finish as important to an Alpine as '*Rosso Corso*' is to a Ferrari. Nothing that any self-respecting start-up car company would not do in a heartbeat, but deeply shocking to a large, traditional carmaker like Renault. Once again, we broke rules, ruffled a lot of feathers, made (insincere) apologies and moved forward ... and the cars looked great.

The Geneva Motor Show, March 2017. I am standing just off-stage, in the 'wings' of the Alpine stand. I am suited and booted, shoes polished and a radio mic' hidden in my waistband. In my slightly sweaty hand are a few flashcards, just in case I fluff my lines. The lights are blazing in my face. I am uncharacteristically nervous. The stand is absolutely packed. It is standing room only – ten deep around the vehicles. People are actually climbing onto the cars on a neighbouring brand's stand to get a better look at ours, stern Swiss security guards motioning them to get down. The expectation is enormous. It's true that Alpine has generated a lot of interest in the media, but I was in no way expecting a crowd this big for a little niche French sports car. We had a bigger crowd than one would normally see for the launch of a new Porsche 911, or the latest Ferrari. I heard the Show announcer say 'And now, please welcome to the stand, Michael van der Sande and David Twohig'. It was time to reveal the car to the world. Michael and I stepped out to do our thing ...

I would be lying if I did not admit that it was a very proud moment. It's rare for an engineer to get the opportunity to present his/her car at a major auto show – that's usually the job of the CEO or the Head of Marketing. Us grimy-fingered geeks usually get to lurk around the back of the show stands, well out of the limelight and hidden from any journalists to whom we might say something silly. But Michael had kindly asked me to join him in presenting the car, and it would be the undoubted highlight of my career. The Show was an unmitigated success. We were inundated on the stand for the week-long duration of the event – by journalists, celebrities, and stars of the motorsport world. I was particularly star-struck in by meeting Gordon Murray of McLaren F1 fame and Ari Vatenen, a particular hero of mine. Mr Murray would famously go on to say some nice things about the car to the media, and later would put his money where his mouth is/was by buying one. He would also make no secret of the fact that he used the car when benchmarking mass control for his multi-million-pound T50 hypercar: a great compliment.

One of the other notable VIP visitors to our stand was the CEO of a rival French car giant – Carlos Tavares himself. He stepped onto our stand to shake our hands under the flash of the photographers capturing a great story – the 'father' of this car, now at the head of the arch-rivals. It was enormous fun to simply hang about the stand, chatting to enthusiasts, potential customers, and just vaguely interested folks, pointing out the tiny areas of attention to detail, especially to save mass, that would prove to be endlessly fascinating to people. The press reaction was very positive. Of course, this was only a 'static' launch – very few people outside the immediate Alpine team had yet driven the car – but the anticipation had been enormous, and the media coverage was enthusiastic. It felt great – a vindication of our work, and a mental break from the overwhelming pressure of managing the technical work while also trying to keep the project financially alive. The Geneva Show was the proverbial double-edged sword – it helped us enormously internally, in that it would now be very difficult to kill the project by pulling the financial plug: that would be almost impossibly embarrassing in the face of such positive press and public reaction. On the other hand, we now had cranked expectations up even higher, and pinned a target on our back. The car looked good – but now it had to deliver, dynamically. As I headed back up the autoroute from Geneva to Paris, I was very conscious that we were far from finished. Time to forget the bright lights of the TV cameras and get back to grinding out this car.

CHAPTER 15

Out of the frying pan ...

The Geneva Show had been a great success, but had the unfortunate side effect – just like the Renault ZOE project five years before – of putting us under the internal spotlight as well as the public gaze. It was almost like waking the giant. Renault – our parent company – had almost forgotten us. We were the bunch of guys in a little industrial unit, led by good old Bernard, then by the tall Dutch guy, with that annoying Irish guy running the engineering team. We were a distraction at best, a mere side-show as the company worked diligently on its 'real' cars. But now things changed. It looked like we might actually succeed in building this car. What was more, the outside world seemed to *really* like it. Geneva had generated an enormous, positive media buzz. Unfortunately, the giant woke up, and decided that we needed 'help.'

Although I was probably less charitable at the time, the efforts of the mothership to 'help' us were undoubtedly well intentioned. The Renault Manufacturing team wanted to apply the standard manufacturing rule book – now pretty much standardised within the Alliance between Nissan and Renault – to our little bespoke plant at Dieppe. Genuinely helpful in many ways, but deeply counter-productive in others. The methods to quality control a Renault Mégane or Nissan Qashqai or a Ford Fiesta rear wing are not the same methods you use to quality control the rear wing of an Aston-Martin or a Ferrari: and we were closer to the latter in terms of production volume, if not price. This was an exceedingly difficult message to get across. Corporate Quality was the same. The team laudably wanted to apply the same quality standards and metrics to Alpine as to any other Renault or Alliance product. For example, let's say that the quality manual would state that 50 samples of a given body panel would have to be precisely and repeatably measured to prove 'capability' of the tool and the press process producing that part. Very sensible when you build 350 cars a shift – 50 parts would represent only an hour's production. Probably overkill when you only intend to build 30 cars a *day*. Suddenly the Alpine team was flooded with 'support' – support that always demanded more documentation, more proof, more samples, more trials, more testing – in short: more.

This sudden emphasis on a very mass-production-oriented corporate interpretation of the word 'Quality' would have been fine if we had been able to relax on the other two corners of the famous Iron Triangle – money (Cost) and time (Delivery). But we were not. The company insisted that we stick to the original budget and that the car hit its planned SOP to the day. Not unreasonable, if we had not had a moving target to hit. But that was becoming more and more difficult. The original idea behind Alpine was that of a true skunkworks – a small, agile team separated from the big, unwieldy mother company, fully empowered to throw away the standard rulebook for mass-market products in order to build something the original company could not have delivered. And, crucially, build it quicker and cheaper that the big company could, even if it dared. But all that was swiftly being forgotten. The company was lifting the lid on the skunkworks and finding that it did not really like the odour of skunk after all ...

The next months were pretty hellish. Every issue – from panel fit and finish to radio reception to a tricky reliability issue with a particular batch of oil pumps – became not only a technical challenge to solve, but a politico-philosophical battle, between Alpine and its concerned parent. As we were required to carry out more and more testing, produce more and more 'standard' documentation, and build more and more expensive pre-series cars, we started to miss programme cost and timing deadlines – at first by a little, then by a lot. And this was not well received. Little mercy was shown as the project went inexorably into the red on the KPIs or Key Performance Indicators so beloved of car companies. The pleasure of the early years of creativity and calculated risk-taking was largely gone, replaced by the dogged trench warfare of any large industrial project. It was vital to not allow the team – or myself – to be destabilised by company politics, however. We had to keep focused on the original objective – affordable agility through light weight. All the other forces of gravity that now wanted to pull us away from that – to add weight, to add cost, to make the car more 'normal,' more Renault-like if you will, had to be resisted. This undoubtedly made us yet more enemies. Even more than ever, it was not a job for the thin-skinned or sensitive. But this obsessive focus was what made cars succeed, and I had already learned that lesson twice over. On Qashqai, it was on achieving the basic 'crossover' concept, and on dogged cost-discipline. On ZOE, it was on delivering the EV technology at a price (cost) point that real people could afford. Somehow, the core Alpine team managed to keep sight of our North Star – low mass at a (relatively) affordable price point – as the corporate storms raged around us.

Leading these teams is often lonely, and no project felt lonelier than the A110. On Qashqai, I always felt cocooned in the warm embrace of Nissan's team-spirit, a genuine, palpable force, not something written on some pious corporate mission statement. Likewise, on ZOE, I always felt the full support of the company behind us – despite the pressure, we knew that the project had full support, from Carlos Ghosn on down, and anything we needed was ours for the asking. Alpine felt very different. But I had access to fail-safe stress relief and therapy tools ... they were sitting in the confidential car park of our anonymous little industrial park in the Renault Sport premises in Les Ulis. All I had to do was get in a prototype or pre-production A110, hit the red START button on that lightweight, floating centre console, and nose it out of the gates and onto the road, to feel the pressure fall away and the love of the project seep back into my bones. The car was magic – it felt truly alive, like an animal, and it had the ability that animals have, to soothe your spirits and make you feel that all this was worthwhile. I particularly remember one drive from Dieppe across country to a plastics supplier deep in the Normandy countryside, to try to solve one of the endless 'capability' issues that we had to document to the satisfaction of our Quality colleagues in *la Grande Maison*. It was midweek, and I was driving a ratty-looking black pre-series A110. It had already had a hard life, was scratched, rattly and unrefined, but had the agility and performance of all A110s. I avoided the French highways or *autoroutes* and threaded it down the back roads, through rolling Norman countryside, probably much faster than I should admit to, the exhaust crackling off the Norman *bocage*, tyres chirping happily through gravelly intersections. The frustrating business of the day and the endless worries of tomorrow faded away, displaced by the endorphins generated by this little pleasure machine. The best meditation I could ever hope for. I parked it up, exhaust and brake discs tinkling merrily, in the supplier's car park, with a friendly security guard briefed to watch it for his life, as I headed for our tense meeting. Back to reality ...

As was the case on ZOE, we ground through the final months of development in a blur of ridiculously long working hours and endless real and fake mini-crises. But finally THE big milestone was passed. SOP, Start of Production – those three little words that cause almost as much trouble as "I love you." There was still a lot of internal tension – our ever-fretful parent company still worried about the car's readiness, still demanded more statistical proof that the 'standard' quality issues were going to be OK. The Alpine team and I continued to argue that it *was* ready – and that measuring the readiness of a low-production, niche, enthusiasts' sports car by the same standards we applied to our mass-production products was simply not the right approach. The arguments were bitter at times – voices were often raised, meetings were so tense that the air needed to be parted with an axe, not a knife. But finally we were there, and the SOP milestone was granted ... with a stack of conditions and a long list of 'homework' to be followed up. I and my team were too exhausted to celebrate. It was just a case of 'it's done.' No satisfaction, no joy, no high-fives or hugs all around – we all just wanted to go home and sleep. Generally, a huge anti-climax.

December, 2017. The long-awaited dynamic Press launch was finally upon us – the Alpine version of the ZOE launch that we'd done in Lisbon five years before. We had hired a boutique hotel near Aix-en-Provence, right in the middle of a beautiful vineyard, with fantastic, narrow, bumpy mountain roads all around, and a little hidden gem of a closed race-circuit called *Le Grand Sambuc*, a few kilometres away. It went as well, or better than we could have ever hoped for. I spent a few days there – again, to accompany the all-important French, German and UK journalists. Especially with the French journalists, we could feel the weight of three decades of waiting for a new Alpine on our shoulders – and on theirs. I was definitely excited as I helped to 'install' the journalists in the cars in the early morning chill, explaining the routes we suggested, talking them through things like Sport modes and the double-clutch gearbox controls. And watching the first cars hurtle past on the circuit with the journalists at the wheel was also exciting, but slightly terrifying – it's always scary to watch someone drive 'your' car at the limit – you are just terrified that someone will have an accident. Thank the automotive gods, we did not have any serious accidents, but the event was not all smooth sailing. We had a sudden cold snap the night before the French journalists came, and both the circuit and the surrounding roads were covered in snow. Michael told me that it was my call, to either pull the event, put the cars on their winter tyres (which would severely compromise the handling) or take the risk that it would clear overnight. I prayed to the weather and the automotive gods, pulled a few strings with the local Mayor to get him to snow-plough some key roads, and took the risk. It paid off … the Mayor came good, the sun burned off the clouds and the thin blanket of snow, and the cars behaved beautifully on their 'summer' tyres. Despite this little scare, the event went like a dream.

Car journalists are pretty hard-bitten, cynical folk. They do not like to give away too much, especially not to the anxious Chief Engineer waiting for them when they pull into the coffee stop or into the pits on the track. But time and again they could not help themselves – they would step out of the car with a huge cheesy grin, or pull a crash-helmet off a sweaty and red-faced head and say things like *"Putain!* That thing is good." They were clearly enjoying it. And, thanks to the internet and the speed of modern journalism, we got almost instant feedback – every night we would eagerly read the verdicts as they appeared on-line – first in French, then in English, then in German. The car was a hit. The press loved it. And they were writing some amazingly complimentary things. The UK press was going crazy over the car. Our hard-to-impress local French journalists had also fallen under its charm – despite the burden of national myths and pride. And possibly most important of all, the German automotive press was as impressed as its UK and French colleagues. Once again, as I had years before with ZOE in Lisbon, I could breathe easy. I knew – at last – that it was going to be okay.

Now that the project was effectively over (or so I thought), I started to think about what was next for me personally. It seemed unlikely that Alpine would launch more new models any time soon – the company was still very nervous about the A110, and wanted to wait and see if it really would succeed before committing to further models. Some of us in the Alpine team pushed back against this as hard as we could, begging the company to have confidence and to push forward with the development of other models, but to no avail. It was clear also that I had put many noses out of joint in the parent company. The fights to make sure that the new Alpine would really be an Alpine and not just a badge-engineered Renault had largely succeeded, but I had definitely pissed off a lot of very senior people. So, my future at Alpine seemed limited, and I was not filled with joy at the prospect of going back to a 'line management' job like my last one at Renault. Yes, I would have Vice-President stripes and hundreds of people working for me, maybe one day Senior VP or even Executive VP stripes, but it would be back to the normal corporate grind – for decades to come. So, when my phone rang one day, displaying the +001 number of a head-hunter based in the US, I half-listened to the gentleman with the California drawl, rather than just giving him a polite 'thanks but no thanks.' He piqued my interest – a new start-up EV company, based in Silicon Valley. Would I be interested in talking to the CEO? I'd always been curious about the US, and about Silicon Valley in particular, and of course the Tesla story had fascinated me as much as anyone else in the car industry, if they are being honest. Fast forward a few weeks, and I had signed up to join this new company, BYTON, as Vice-President and Chief Vehicle Engineer. I was 48, had painted myself – albeit knowingly – into a bit of a career corner at Renault, so it was now or never for a move like this. I handed my letter of resignation to my boss, who was surprised but understanding. Thierry Bolloré – then head of Renault, having replaced Carlos Tavares – was kind enough to find an hour in his very busy schedule the next day to try to talk me out of my decision, but it was too late. We had a great conversation; he shook my hand warmly and wished me the best of luck.

Having handed in my letter of resignation, I was looking forward to quietly working my six weeks 'notice,' to ensure a smooth transition and to make sure there were no hiccups before Start of Sales, planned in a few weeks. I was looking forward to a relaxing few weeks – the car-park at Dieppe was slowly but steadily filling with beautiful Alpine-blue A110s, ready and waiting to be loaded on the car transporters to ship them to the brand-new Alpine showrooms all over Europe. I was not too worried – of course the Quality folks and the Renault Manufacturing 'experts' were still bitching about some of the minor issues, but I did not see any major issues ahead. How wrong I was ...

Monday, 29th January, 2018. I was in the Renault Sport/Alpine Engineering management Monday morning meeting, half-listening to the routine tasks-of-the-week outline and starting to crave my mid-morning caffeine hit once the meeting was over. My laptop binged with an instant message from Michel Fumex: "You need to come out of the meeting. Big problem."

"Can't it wait?" I typed back, discreetly.

"No. Come out now." I was rather annoyed – the team were big boys, could they not wait a half-hour? I already visualised my morning coffee evaporating into whatever this latest panic was. But the next message from Michel made the hairs stand up on the back of my neck – Michel was a serious person, not prone to practical jokes:

"It's the *Top Gear* car. It caught fire."

Rewind. Since the successful launch in Aix-en-Provence, our PR team had been lending pre-production 'Press' cars out to the most important media outlets in Europe. A few days before, I had approved the release of one of the cars to Renault UK, to provide to the UK's *Top Gear* TV program. Although viewing figures had fallen a little from when this ultra-popular program had starred Jeremy Clarkson, Richard Hammond and James May, the latest incarnation of the show was still wildly popular. Hence, it was impossible to refuse to give them the car, despite their reputation for being pretty harsh on anything that did not perform up to their standards. Anyway, I was confident – the car was damn good, we had nothing to hide. I was also a big fan of *Top Gear*'s main presenter and expert test-driver, Chris Harris. I'd always enjoyed his writing and thought he was a great asset to the show: genuinely passionate about cars, straight-talking and spectacular behind the wheel. I was sure that he'd appreciate the handling of the car and that he'd 'get' the whole low-weight/low-power ethos. So, the paperwork was signed, and the car was shipped off to the UK, where it would do some photography before being transported to the South of France, where *Top Gear* wanted to film a little piece during the Monte Carlo rally, driving the car on a stretch of one of the rally stages in between the 'real' rally cars. Perfect ... a beautiful little link to our rally heritage, some great scenery. Chris Harris would hopefully rave about the car. What's the worst that could happen?

Well, when I stepped out of the meeting to meet a welcome party comprising a pale-looking Michel and a few other team members, I found out. The car had caught fire. I quickly called the Renault UK PR 'minders' who were on site and got the details. It could not have been much worse. Not only had the car caught fire, but it had caught fire *on camera*. The presenters – Chris Harris in the driver's seat, and Eddie Jordan in the passenger bucket – had had to jump out of it while flames licked around the sills. Unfortunately, none of the chase/camera cars had a fire extinguisher at hand, and the car had literally burned to the ground – more of that later. Worse still, the smoking remains were still on the side of a public road, with passers-by, all armed with mobile phones, hence cameras. The only good news – and it was an enormous relief to hear this – was that nobody had been hurt.

I immediately started to work through the battery of my mobile phone.

My memories of that day are of constantly wishing I were there myself (*genba* – remember Ogawa-san?) to gather the facts. But it was almost 10 hours away by road from our base near Paris. Failing being there myself, I got hold of some of our press people on the scene. First priority – get what remained of the car covered. The team found a tarpaulin and later a sort of improvised marquee-tent to hide it from prying eyes. Next, I needed to know exactly what had happened, in detail. I'd had some experience of analysing vehicle fires in previous roles, and I knew that it was very like a crime scene investigation – every detail counted, and time was critical. I called Jeremy Townsend, then head of our UK Press Office. I told him that I needed Chris Harris' personal mobile phone number. This broke any number of rules. First, the BBC does not simply give out their star journalists' phone numbers. Second, Chief Vehicle Engineers do not – ever – speak to the Press unchaperoned, especially when said Member of the Press has just narrowly avoided being char-grilled in one of our products. But Jeremy is a star. He just said "Leave it with me" in a typically British understated way, and 20 minutes later I got a text message, with Mr Harris' number. The phone rang a couple of times, and then a voice I knew very well from the TV said "Chris Harris. Hello?"

"Mr Harris, sorry to call you directly. My name is David Twohig and I'm Chief Vehicle Engineer at Alpine. You can probably guess why I'm calling you."

"Yeah, I probably can!" Oh dear. Slightly truculent-sounding. Fair enough – I would probably be more than truculent if I'd just leapt out of a burning car, and the guy who'd designed it called to ask how my day was going. But luckily I got off on the right foot.

"Look, first thing to ask you – are you and Mr Jordan both okay?"

I heard his voice relax a little as he said yes, no worries, both were hale and hearty. We then had a great conversation. I asked him if he could take me through everything that happened. I told him I wanted to know the time he got up, what he had for breakfast that day and exactly how the situation developed from there. He laughed, and took me through the events in remarkable detail and with amazing precision. I should not have been surprised, as Chris is after all a professional racing driver as well as a journalist, and hence has those fighter-pilot cool nerves. But his recall *was* truly remarkable. I was furiously scribbling down almost every word he said, and afterwards I was able to cross-check both the sequence and the timing of those events with great accuracy. I sincerely thanked him, and he very graciously told me to call him back any time that I needed more information – he'd be glad to help. I did actually take him up on that once or twice, and every time I called, he was a true gentleman.

Next, more rule-breaking. Jeremy T again. "Okay Jeremy, I need the actual video that the BBC shot. Every bloody frame. And I need it pronto." Even Jeremy gulped a bit this time before his usual phlegmatic "Leave it with me." This request was harder, and 30 minutes later he called me back. The BBC never released footage of high-profile shows like *Top Gear* before the broadcast date. But it would make an exception this time, and would release it to *me* – and me only – on a secure server with a password that only I would have access to. Fine. Done. Sign me up. Oh yeah, Jeremy explained that there was one other thing. They had to have a car by 26th February, to finish shooting, and to film the famous timed lap of the *Top Gear* test track back in the UK, with the white race-suited 'Stig' at the wheel. If we could not supply a car by then, well, they would simply broadcast the only footage they had 'in the can' – ie, the video of the car engulfed in bright orange flame. I felt a cold finger run down my spine. I knew how spectacular that would look on the TV, how many millions would watch it. It would be the death of the car and of the Alpine brand, for good this time. The Beeb[43] was gently but firmly telling us that we had three weeks to sort our shit out and to give them a car that worked – or else. Great.

Next, I had to issue a 'stop order' for every car in existence. I'd only ever done this once before, but I immediately drafted a memo to every department and individual that had the keys of an A110, including the production plant at Dieppe: stop now, park the car, don't move it until further instruction. No explanation of why – we could not afford for this to leak. All hell broke loose – my e-mail inbox exploded with questions that I could not answer. Ignore them. Next task – get the *Top Gear* car, or what was left of it, back to our workshops, ASAP. This took some organisation. The car had literally burned not just *to* the ground but *into* the ground – the extreme temperatures had actually melted the surface of the asphalt, and fused the car's undertray and axles into the road. So, the car could not just be towed or dragged onto a normal flatbed car transporter – it needed to be lifted, complete with a 5m (~16ft) by 2m (~6ft 6in) slab of melted tarmac – *onto* a flatbed, using a hydraulic arm or 'hi-ab.' It took some desperate phoning around to find a local recovery operator who had such a truck and could make it up the narrow winding Alpine roads to where the wreckage lay. I spoke to the truck owner myself. By now, it was already late afternoon. I told him I needed the car to be in Paris, by 7am the next morning. I wanted to get the car into our secure workshops before 8am, when most of the technicians in the building started work. This was still very secretive – nobody knew, except the few people at the scene, that our car had caught fire. The fewer people that knew, the better, before we knew what the cause was. The truck owner told me the best he could do was maybe lunch-time – his driver was up against his driving time-limit, and would have to stop to sleep. Luckily, I had been in France long enough to know that the word '*non*' is just an invitation to discuss, cajole and negotiate. I asked him how much it would cost to hire a

---

[43] *Affectionate British nickname for the BBC.*

second driver, to take over when the first guy would time out, and get the car to Paris, at all costs, early the next morning. Pause. 6000 euro. This was extortionate, of course. 1000 euro would probably be reasonable, 2000 euro would be expensive. I did not hesitate. Done. Next morning, I met the truck – with its two sleepy-looking drivers, in the car park of our engineering centre. It was cold and dark, with a dank February drizzle. It was too early for the Finance department to be there, and in any case, I hadn't followed anything remotely resembling a procurement procedure here. I wrote the guy a personal cheque for the six grand on the hood of his truck. I gave him my business card and told him he knew where to find me if it bounced. He shook my hand and pocketed it. He then opened the tarpaulin on the back of his low-loader, swung the crane and dropped a ton of melted tarmac, charred aluminium and burned wreckage on a large rolling pallet that we had pulled into place behind the truck. Michel, myself and a few of the guys sadly wheeled it through the workshop area and into a confidential 'box' in which we would carry out the investigation. The whole workshop suddenly stank – the air in the building immediately filled with the acrid, throat-burning stink of melted metal, rubber and plastics. This was now the centre of the 'crime scene.'

The next two weeks were simultaneously the most exhausting, and the most fulfilling, of my professional career. Everything seemed to come together in a non-stop marathon of analysis – my now long-ago training as an engineer, the step-by-step calm analysis of the facts, the '5 Whys' that Nissan had taught me long before, the intimate knowledge of almost every part of this car, came together in understanding the causes, and (more importantly) fixing them. All carried out as the clock ticked inexorably in my mind – not only the internal Start of Sales date, now postponed, but the 26th February external deadline imposed by the BBC: the date after which they would broadcast images of my career, reputation, and the future of this brand I loved, quite literally going up in flames.

In hindsight, it was fantastic – true, pure engineering. Nobody bothered me with requests to sign off their expenses, or to approve their vacation. Nobody wanted to have a one-to-one with me to complain about not being promoted, or to ask about their career prospects. Nobody wanted to tell me how to handle this situation, or even suggested the best way to do so – nobody had ever been presented with this. It was a huge, loudly ticking time-bomb and nobody wanted to be remotely close to it this time. It was wonderful. Even my recent sparring partners in the Renault quality department were helpful. The situation was not managed by the 'normal' quality teams, but by Renault's corporate crisis management team – the team that handled major issues like fatal accidents, major recalls etc. And they were very clear and straightforward to deal with – their only condition was that no car was to be supplied to the BBC until I had found and fixed the issue and that all the cars already built had been fixed, so that we could declare Start of Sales. Clear. Almost impossibly difficult, but very clear. Apart from that, they just wanted a daily report on progress, wished me the very best of luck, stepped aside and left us to get on with it, in the way that you let a guy defusing a very large and unstable bomb get on with it. You tend not to look over his shoulder. Perfect. To boot, I now had privileged access to everything – all of the Renault Group's considerable resources. Anything I needed was just a phone call away: no more griping about slightly exceeding development budgets now.

I virtually lived at the office for the next two weeks. The analysis was based on a Failure Tree Analysis done in the old-school way – dozens of paper post-its, stuck to the opaque glass rear wall of my office – no copies, no photographs. Most of my time was spent – at least for the first three days or so – sifting through the wreckage. Renault had a vehicle fire analysis expert, a specialist in investigating vehicles that had burned out, to establish if the fault was a technical issue, or (as is often the case) arson for insurance reasons. He had been assigned to my team for the duration. He and I were the only ones authorised to touch the wreck. For three days we sifted through it, not with tools, but with our gloved fingers, moving hundreds of kilos of ash aside, layer by layer, cataloguing every identifiable part of the car as we found it. As we turned up a part, we would note its position and depth, just like an archaeological dig, photograph the part in place, then pass it to a few trusted helpers, who bagged it and stored it carefully on metal racks.

The fire had melted everything that was not ferro-metallic (iron or steel) or glass. The heat had been enough to melt aluminium – in this very aluminium-intensive car, all that was left was the flaky, greyish crumbly ash that is left when this metal is heated well above its melting point of 660°C. But there were clues – all the (steel) nuts and bolts were there, parts of circuit boards were intact, their silicon chips largely intact. We could find the glass lenses of the onboard cameras that the *Top Gear* team had left, hose clips, steel electrical connectors. For three days we sieved through the debris, backs aching from bending over the pallet, our blue Alpine overalls stained black, faces grimy like old photos of coal miners. When we would take a drink break to wash the taste of ash out of our mouths, we stank. Folks at the water coolers nodded and did not ask 'how's it going?' – our blank facial expressions told the story. When I went home exhausted every night, the water in the shower ran black as the ash washed out of my hair and off my skin.

When I was not being a coal-miner-archaeologist, I was obsessively watching the highly confidential and password-protected *Top Gear* footage on my computer, over and over, again and again, freeze-framing it at the key points, inching it forward frame by frame, listening to Chris Harris and Eddie Jordan's voiceover, starting in their usual laddish car-journalist banter, then turning serious as Chris said "I'm losing power. Power gone ... Fuck, it's on fire. Get out, Eddie! Now!" And slowly, through the physical evidence and the video record, we started to figure it out. Fire is deceptively simple to analyse. You need three things – fuel, oxygen and a source of ignition. In our case, the second did not need analysis – the car burned in the open air, and air is 21% oxygen. Done. It was relatively easy to find the fuel ... but harder to prove it. You are going to say – duh, the *fuel* was the fuel, idiot! And of course it was, eventually, the half-full petrol tank that provided the fuel for a fire so intense that it destroyed even metals.

But we were looking for the *initial* fuel – the source of the fire. And cars are full of stuff that can burn, besides their petrol or diesel fuels – oil, air-conditioning gases, brake fluid, greases, even engine coolant can burn if the conditions are right. But we were able to establish, by the way the power cut out and the precise timing on the video time-stamp of the power loss, and of the first flames licking around the car, that it was indeed a fuel leak, and highly probably from the fuel lines leading from the front-mounted fuel tank to the mid-mounted engine. But 'highly probably' was not good enough – we needed proof. Physical proof. And unbelievably, we found it.

We were able to simulate the most likely aggressive element that might have damaged the fuel line – a fingernail-sized 8mm nut on the front of the engine. As the engine moved around on its mountings ... which it does as the car is 'drifted' through a corner, as Chris had been doing, we could show that it might (just *might*) come in contact with this particular nut. And we had actually *found* this nut, when we had sifted patiently through the wreckage. The aluminium engine block had melted, but the nut had been found in its cocoon of ash, like a Pompeii corpse, and had been duly bagged up and labelled. And now I could rely on the technical muscle of a huge industrial corporation like Renault – I was able to ask the central materials labs to analyse it with electron microscopy and chemical spectroscopy. I know this sounds like a bad episode of NCIS, but I promise you it's true – we were able to find microscopic abrasion of the nut (yes, a hard steel nut) where the high-strength plastic of the fuel line had rubbed against it, slightly rounding and polishing the metal crystals. Even more unbelievably, the chemists were able to ascertain that liquid fuel had been spilled on the nut before the fire, ie a fuel leak had happened here, *before* the whole lot went up. So, we had our source of fuel – and it was pretty easy to redesign the fuel line routing to move it further away from the offending nut. Just to add braces to the trouser-belt, we took the double precaution of also protecting it with a very tough synthetic mesh, somewhat like chain-mail armour for plastic parts.

Ignition was easier – once fuel is sprayed out under high pressure into the oxygen-rich air of an engine bay, it's going to find a source of ignition sooner or later – either a hot exhaust manifold (exhaust manifolds can heat up to over 800°C – well over what's known as the 'auto-ignition' or self-ignition temperature of petrol – about 280°C) or micro-sparks from an alternator's brushes, or several other potential fire-lighters. It took a little more computer simulation and physical test work to track these down, but eventually we had found – and demonstrated beyond any reasonable doubt – the sources of both fuel and ignition.

Time was running out – we had just a few days to go now before the internal Start of Sales and external 'BBC' deadlines. Finally, we managed to retrofit all the cars with the fixes, and carry out extensive bench and vehicle testing to prove that those fixes worked. I reported back to the quality crisis management cell, who had, true to their word, let me get on with it. They were happy with the results. We were good to go. We had one day to get a car to *Top Gear* in the UK. One of the now-fully-production-representative cars was shipped to the show's 'secret' test track in southern England. When it arrived, the windscreen was broken. Just great. More favours. I called a couple of our technicians and asked them to grab a van, drive to Dieppe, pick up a new screen and keep driving ... right across the Channel Tunnel and into the UK to fit the screen, that very night. I myself jumped on a flight to Heathrow and hotfooted it to the *Top Gear* test track that night – partly to support the two guys, whose English was pretty ropey, but mainly to hand the car over to the *Top Gear* presenters myself. I thought it was the least we could do.

And that was a very cool way to end my involvement with the project – I drove this bright blue little car into the paddock (car park, really) at the BBC's test-track/studio, slightly intimidated by the various people running around busily with camera equipment, head-sets and clipboards. I admit also to being slightly star-stuck when Matt LeBlanc or 'Joey from *Friends*' walked by, flashed me that famous smile and said, "Nice car, man!" We were in good company – I parked up next to the latest Aston-Martin Vantage and Porsche 911 GT2 RS. We were down about a hundred grand and several hundred horsepower in this company. My Alpine-branded jacket and sheepish demeanour must have attracted attention, as Chris Harris himself walked up to me and said "You must be Dave. Did you suss it out in the end?" I was happy to confirm that we had – without boring him with the literally grubby details – and we had a very nice chat. The Stig did his thing – very impressively – and blasted the car efficiently around the *Top Gear* test track. It looked great, and did a very respectable 1' 22.9" lap – nothing to worry the 911 GT2 RS, but I was more than happy with that time.

And all ended well – the BBC played very fair with us when the episode was eventually broadcast on 1st April 2018. They said some very nice things about the car, cracked a few gentle jokes, and did end up showing a – very brief – image of the car in flames. One could not blame them – it was simply too good a story, and too spectacular an image, *not* to use. But they enthused about the car, stressed that it had been a pre-production model, and read out word for word the very bland – but 100% accurate – Press statement that we had carefully prepared, basically saying that the issue had been solved for all production cars. In the end, the piece probably did more good than harm to the car in the UK and abroad. And again, with hindsight, it provided me with a highlight of my career – those two weeks sifting through ashes, staring obsessively at the unreleased *Top Gear* video and sticking post-its on my office wall, were probably the best and purest 'engineering' I ever did, or may ever do.

My last days in Alpine were so sweet that they were almost bitter. The whole *Top Gear* drama was behind us, and I could enjoy my last few days working for Alpine, for Groupe Renault and for the Alliance that had employed me ever since I was a snot-nosed kid engineer turning up at NTCE Cranfield with a head full of nothing useful. Nobody seemed to hold a grudge against me for leaving – and I guess more than a few were quietly glad to see the back of me! The last few days passed in a pleasant round of handing over documents and files to my replacement – ironically and very fittingly, Jean-Pascal Dauce, the very guy who had laid such excellent foundations for the project and who had handed it over to *me* a few years before. In some ways it was like handing him back his baby after I had fostered it for a while. The car was still getting rave reviews in the motoring Press, and the first of what was to become a flood of awards were starting to come in.

The team was gracious enough to organise a very nice leaving party for me – very French, with cafés and pâtisseries, and speeches by various folks including Thierry Bolloré, who had very kindly made the time to drive over from Renault company headquarters to wish me good luck. The team gave me some truly priceless gifts as well – a numbered plaque from one of the cars, a hand-painted artwork of the car, and a 2 x 1 metre picture of the A110 on an aluminium plaque, part of the Geneva show stand, and signed on the back by the whole team. I still have them all, and will never part with them. A day or two after the party, I left my security badge, computer, company phone and car keys at the front desk of the Renault Sport/Alpine technical offices and walked out the door for the last time as an employee. Two days later I was on a plane to take up my new role in BYTON – and the next adventures that awaited me in Germany, California, China, Korea and Dubai. But that, as they say, is another story ...

Epilogue and Acknowledgements

EPILOGUE

Epilogues usually bring the reader up to date with the subsequent story of the principal characters of a book. Hopefully, the characters that emerge from this book have been the cars. So here is what became of P32L/Qashqai, B10/ZOE and, of course, the Alpine A110.

Qashqai was launched in February 2007 and sold like the proverbial hotcakes. Suzuki-san's ambition of maybe achieving 130,000 units sold in the peak year was achieved in months. The car went on to sell well over 200,000 units a year for many years in a row, peaking at over 340,000 units built in a single year. It was mildly face-lifted in 2010, then replaced by the 2nd Generation Qashqai in 2013. This model had a substantially new upper body, but it sat on largely the same basic platform as my team had sketched out in Japan, way back in 2003. This platform also provided the underpinnings for vehicles such as the Nissan Rogue Sport, X-Trail and Renault Koleos. In 2021, Nissan launched the 3rd Generation Qashqai, which is essentially a whole new vehicle. To date, Qashqai has sold well over 3.3 million vehicles, and is still going strong. It has been sold worldwide, and remains a key pillar of Nissan's profitability. The cheap pun 'Cash Cow' has probably been over-used, but if the shoe fits ...?

What are the reasons for this immense success? A huge dose of luck, to be honest: it was the right product at the right time – a good-looking, affordable, reliable crossover, launched into a market just starting to get bored with cookie-cutter hatchbacks and MPVs. Stéphane's restrained design and Suzuki-san's remarkable restraint in product planning were keys to this success, as was my team's discipline in execution, and specifically, cost-control. Qashqai has won a list of awards over the years that is simply too long to detail here. Far more importantly, it's been the source of stable employment for tens of thousands of staff at Nissan in the UK, and in suppliers and supporting companies all over the world for over a decade. Those who were involved in its genesis are quietly but deeply proud of that.

Unlike the instant-success Qashqai, ZOE was a slow burner. Initial sales after its launch in spring of 2013 were disappointing, but as the world became more concerned with the impact of the internal combustion engine on the environment, and as EVs gradually became 'cooler,' sales slowly but steadily ramped up. Over 300,000 ZOEs have been sold at the time of writing, and it was the best-selling all-electric vehicle in Europe for many years – holding off competition from the likes of BMW, VW and even from those cool Californians, Tesla. Its dominant market position was maintained by clever lifecycle management of the basic platform by the Renault engineers, improving the traction motor shortly after launch, and progressively updating the battery, as lithium-ion battery energy density steadily improved over the years. It was not the first practical, affordable EV – that honour must go to the Nissan Leaf – but kindly indulge my admittedly-biased opinion as I claim that it was the first affordable and *cute* EV, thanks to the great design work by Geoff Gardiner and Jean Smeriva. Its looks have not dated, and many millions of people have had their first experience of EVs at the wheel of their 'little mouse.' A new generation of the ZOE, but still with the same technical DNA, is still in production at the time of writing, and shows no signs of sales dropping off.

Finally, the Alpine A110. Probably the most acclaimed of the three vehicles, but the least commercially successful. To say that the A110 was well received by the automotive press is somewhat of an understatement. It is one of the very few cars to score a perfect 5/5 or 10/10 in many prestigious magazine tests. Listing the awards that it has won would be a long and tedious exercise – in short, it's won everything there is to win in the way of 'best sports car' awards. It has a cult following among enthusiasts, even in territories where it is not even sold, like the US. But despite the acclaim, it has been a lukewarm commercial success. It has sold, at the time of writing, approximately 9000 units in its four-year production life – admittedly more than the original A110 *Berlinette* sold in its marathon 16-year production life, but Porsche has sold as many Caymans in a single year, to put that figure in perspective. The car is probably condemned to forever be a rare, niche product, appreciated by true sports car enthusiasts, but largely unknown to the general public. Ironically, its true impact may well be the effect it will have on *other* manufacturers' products. Its low-mass, low-power ethos was genuinely revolutionary in a market trending, seemingly inexorably, towards ever-heavier, ever-more powerful sports cars. I am very confident that every sports car maker has quietly purchased an Alpine A110, to try to understand how it succeeded in weighing so little. Some have been open in their admiration of the technical approach – Gordon Murray, for example. So hopefully it will have an influence on the sports car industry and technology far beyond its actual sales figures. At the time of writing, a new Management team in Renault Group has re-launched investment in the brand, and finally committed to a new range of Alpine vehicles. Great news for the Alpine fans and enthusiasts. Hopefully 'Monsieur Jean' Rédélé's vision will continue well into the 21st century.

ACKNOWLEDGEMENTS

Great thanks to Rod Grainger, Becky Martin, and all the team at Veloce for their support throughout the process of writing this book. Thanks also to Brian Curran and Tim Dempsey for their early input, encouragement, and advice. Dan Prosser and Andrew Frankel at *The Intercooler* publication also provided a lot of encouragement, as well as also taking a risk by publishing work by an old soldier of the car industry, but a neophyte in the publishing world.

Fiona Desmond, Kathleen Murphy, John Murphy, Trish Twohig and Steve Matthews read early drafts, picked up errors and provided valuable advice. The good bits are thanks to them – the boring bits are all my own work.

The three Chief Designers of the cars described also forgave my impatience and spikiness during the years we worked together, and kindly created the original sketches for the book's cover, despite immense demands on their time. Thanks so much, Stéphane, Geoff and Antony.

Thanks also to some experienced professionals – Richard Porter, Kevin Dwyer, Liam McCann, Bernard Ollivier, Jeremy Townsend and Neil Briscoe – all of whom provided valuable advice to a beginner: advice that was free of charge, but full of wisdom. Folks like this really do restore one's faith in human nature.

My old colleagues in the communications and photo archive services at Nissan and Renault were very generous with their time in digging out many of the images that illustrate the book. Thanks to Adrian Smart, Simon Fraser, Peter Brown, Simon Bottomley, Matthew Walker, Dominic Vizor, Hélène Bocquet, and Patricia Esposito in particular.

And, like all engineers and authors who are probably unhealthily obsessed with their industry and projects, thanks are most certainly due to my unbelievably patient wife, Cooleen. She's endured my car industry tales for years. Now I've at least got some of them out of my system.

Finally, thanks to all the generations of car engineers that came before me, and more particularly to the engineers and managers that trained me. I've mentioned some but by no means all of them in the above text. The list of those who patiently bore with my errors and showed me how to avoid them in future is simply too long to commit to paper. But I humbly thank them all – and if this book encourages just one young engineer or potential engineer to stand on the shoulders of these giants, well, my work here will be done.

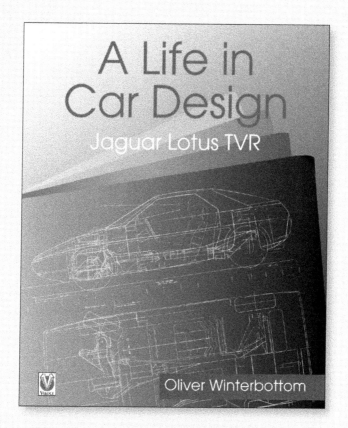

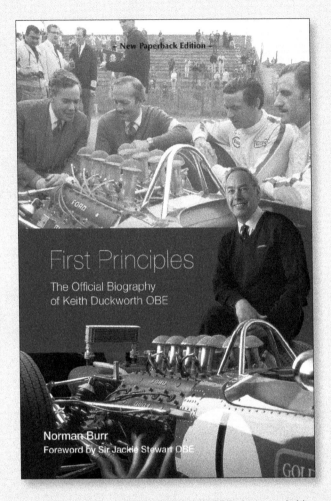

This book chronicles the life of Keith Duckworth OBE, the remarkable engineer, co-founder of Cosworth Engineering and creator of the most successful F1 engine of all time, the DFV. This is a rounded look at the life and work of the man – work which included significant contributions to aviation, motorcycling, and powerboating.

ISBN: 978-1-787111-03-5
Paperback • 23.2x15.5cm • 352 pages • 200 pictures

For more information and price details, visit our website at www.veloce.co.uk
email: info@veloce.co.uk • Tel: +44(0)1305 260068

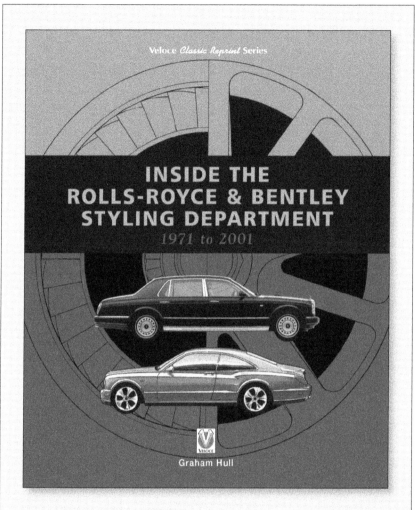

A Veloce Classic Reprint.
The unique and personal account of young designer's journey after joining that most prestigious of marques, Rolls-Royce. Sometimes eccentric, often humorous, the workings of this uniquely British institution during a period of immense change are described in detail.

ISBN: 978-1-787115-47-7
Paperback • 25x20.7cm • 176 pages • 100 colour and b&w pictures

For more information and price details, visit our website at www.veloce.co.uk
email: info@veloce.co.uk • Tel: +44(0)1305 260068

Index